Composite structures
of steel and concrete

Composite Structures of Steel and Concrete

Beams, slabs, columns, and frames for buildings

Third Edition

R.P. JOHNSON

MA, FREng, FICE, FIStructE
Emeritus Professor of Civil Engineering
University of Warwick

Blackwell
Publishing

© 2004 R.P. Johnson
© 1994 Blackwell Scientific Publications
© 1975 Constructional Steel Research and
Development Organisation

Editorial offices:
Blackwell Publishing Ltd, 9600 Garsington Road,
Oxford OX4 2DQ, UK
   Tel: +44 (0)1865 776868
Blackwell Publishing Inc., 350 Main Street, Malden,
MA 02148-5020, USA
   Tel: +1 781 388 8250
Blackwell Publishing Asia Pty Ltd, 550 Swanston
Street, Carlton, Victoria 3053, Australia
   Tel: +61 (0)3 8359 1011

First published by Crosby Lockwood Staples 1975
Paperback edition published by Granada Publishing
1982
Reprinted 1984
Second Edition published by Blackwell Scientific
Publications 1994
Third Edition published 2004

Library of Congress Cataloging-in-Publication Data
Johnson, R.P. (Roger Paul)
   Composite structures of steel and concrete :
beams, slabs, columns, and frames for buildings /
R.P. Johnson.
      p. cm.
   Includes bibliographical references and index.
   ISBN 1-4051-0035-4 (hardback : alk. paper)
   1. Composite construction.   2. Building,
Iron and steel.   3. Concrete construction.
I. Title.

   TA664.J63   2004
   624.1′821–dc22

                                    2004000841

ISBN 1-4051-0035-4

A catalogue record for this title is available from the
British Library

Set in 10/13pt Times
by Graphicraft Limited, Hong Kong
Printed and bound in India
by Replika Press Pvt. Ltd, Kundli 131028

For further information on Blackwell Publishing,
visit our website:
www.thatconstructionsite.com

Contents

Preface

This volume provides an introduction to the theory and design of composite structures of steel and concrete. Readers are assumed to be familiar with the elastic and plastic theories for the analysis for bending and shear of cross-sections of beams and columns of a single material, such as structural steel, and to have some knowledge of reinforced concrete. No previous knowledge is assumed of the concept of shear connection within a member composed of concrete and structural steel, nor of the use of profiled steel sheeting in composite slabs. Shear connection is covered in depth in Chapter 2 and Appendix A, and the principal types of composite member in Chapters 3, 4 and 5.

All material of a fundamental nature that is applicable to structures for both buildings and bridges is included, plus more detailed information and a fully worked example relating to buildings. The design methods are illustrated by calculations. For this purpose a single problem, or variations of it, has been used throughout the volume. The reader will find that the dimensions for this structure, its loadings, and the strengths of the materials soon remain in the memory. The design is not optimal, because one object here has been to encounter a wide range of design problems, whereas in practice one seeks to avoid them.

This volume is intended for undergraduate and graduate students, for university teachers, and for engineers in professional practice who seek familiarity with composite structures. Most readers will seek to develop the skills needed both to design new structures and to predict the behaviour of existing ones. This is now always done using guidance from a code of practice. The British code for composite beams, BS 5950:Part 3, Section 3.1, is associated with BS 5950:Part 1 for steel structures and BS 8100 for concrete structures. These are all being superseded by the new European codes ('Eurocodes'), and will be withdrawn within a few years. The Eurocodes are being published by the standards institutions for most European countries as EN 1990 to EN 1999, each of which has several Parts. These have been available as ENV (preliminary) codes for several years.

In the UK, their numbers are BS EN 1990, etc., and in Germany (for example) DIN EN 1990, etc. Each code includes a National Annex, for use for design of structures to be built in the country concerned. Apart from these annexes and the language used, the codes will be identical in all countries that are members of the European Committee for Standardization, CEN.

The Eurocode for composite structures, EN 1994, is based on recent research and current practice, particularly that of Western Europe. It has much in common with the latest national codes in this region, but its scope is far wider. It has many cross-references to other Eurocodes, particularly:

- EN 1990, Basis of Structural Design,
- EN 1991, Actions on Structures,
- EN 1992, Design of Concrete Structures and
- EN 1993, Design of Steel Structures.

All the design methods explained and used in this volume are those of the Eurocodes. The worked example, a multi-storey framed structure for a building, includes design for resistance to fire. Foundations are not included.

The Eurocodes refer to other European (EN) and International (ISO) standards, for subjects such as products made from steel and execution. 'Execution' is an example of a word used in Eurocodes with a particular meaning, which is replacing the word in current usage, construction. Other examples will be explained as they occur.

Some of these standards may not yet be widely available, so this volume is self-contained. Readers do not need access to any of them; and should not assume that the worked examples here are fully in accordance with the Eurocodes as implemented in any particular country. This is because Eurocodes give only 'recommended' values for some numerical values, especially the γ and ψ factors. The recommended values, which are used here, are subject to revision in National Annexes. However, very few of them are being changed.

Engineers who need to use a Eurocode in professional practice should also consult the relevant Designers' Guide. These are being published in the UK for each Eurocode, and are suitable only for use with the code and those to which it refers. They are essentially commentaries on a clause-by-clause basis, and start from a higher level of prior knowledge than is assumed here. The Guide to EN 1994-1-1, *Design of composite steel and concrete structures – General rules and rules for buildings* is consistent with this book, being written by the present author and D. Anderson. Corresponding publications in other languages are appearing, each relating the Eurocodes to the national codes of the country concerned.

The previous edition of this volume was based on the ENV Eurocodes. The many changes made in the EN versions have led to extensive revision and a complete re-working of the examples.

The author has for several decades shared the challenge of drafting the General, Buildings and Bridges parts of EN 1994 with other members of multi-national committees, particularly Henri Mathieu, Karlheinz Roik, Jan Stark, Gerhard Hanswille, Bernt Johansson, Jean-Paul Lebet, Joel Raoul, Basil Kolias and David Anderson. The substantial contributions made by these friends and colleagues to the author's understanding of the subject are gratefully acknowledged. However, responsibility for what is presented here rests with the writer, who would be glad to be informed of any errors that may be found.

Thanks are due also to the School of Engineering, University of Warwick, for facilities provided, and most of all to the writer's wife Diana, for her unfailing support.

R.P. Johnson

Cover photograph shows composite decking prior to concreting (courtesy of The Steel Construction Institute).

Symbols, terminology and units

The symbols used in this volume are, wherever possible, the same as those in EN 1994 and in the Designers' Guide to EN 1994-1-1. They are based on ISO 3898:1987, *Bases for design of structures – Notation – General symbols*. They are more consistent than those used in the British codes, and more informative. For example, in design one often compares an applied ultimate bending moment (an 'action effect' or 'effect of action') with a bending resistance, since the former must not exceed the latter. This is written

$$M_{Ed} \leq M_{Rd}$$

where the subscripts E, d and R mean 'effect of action', 'design' and 'resistance', respectively.

For longitudinal shear, the following should be noted:

- v, a shear stress (shear force per unit area), with τ used for a vertical shear stress;
- v_L, a shear force per unit length of member, known as 'shear flow';
- V, a shear force (used also for a vertical shear force).

For subscripts, the presence of three types of steel leads to the use of 's' for reinforcement, 'a' (from the French 'acier') for structural steel, and 'p' or 'ap' for profiled steel sheeting. Another key subscript is k, as in

$$M_{Ed} = \gamma_F M_{Ek}$$

Here, the partial factor γ_F is applied to a *characteristic* bending action effect to obtain a *design* value, for use in a verification for an ultimate limit state. Thus 'k' implies that a partial factor (γ) has not been applied, and 'd' implies that it has been. This distinction is made for actions and resistances, as well as for the action effect shown here.

Other important subscripts are:

- c or C for 'concrete';
- v or V, meaning 'related to vertical or longitudinal shear'.

Terminology

The word 'resistance' replaces the widely-used 'strength', which is reserved for a property of a material or component, such as a bolt.

A useful distinction is made in most Eurocodes, and in this volume, between 'resistance' and 'capacity'. The words correspond respectively to two of the three fundamental concepts of the theory of structures, equilibrium and compatibility (the third being the properties of the material). The definition of a resistance includes a unit of force, such as kN, while that of a 'capacity' does not. A capacity is typically a displacement, strain, curvature or rotation.

Cartesian axes
In the Eurocodes, x is an axis along a member. A major-axis bending moment M_y acts about the y axis, and M_z is a minor-axis moment. This differs from current practice in the UK, where the major and minor axes are xx and yy, respectively.

Units
The SI system is used. A minor inconsistency is the unit for stress, where both N/mm^2 and MPa (megapascal) are found in the codes. Similarly, kN/mm^2 corresponds to GPa (gigapascal). The unit for a coefficient of thermal expansion may be given as 'per °C' or as 'K^{-1}', where K means kelvin, the unit for the absolute temperature scale.

Symbols

The list of symbols in EN 1994-1-1 extends over eight pages, and does not include many symbols in clauses of other Eurocodes to which it refers. The list can be shortened by separation of main symbols from subscripts. In this book, commonly-used symbols are listed here in that format. Rarely-used symbols are defined where they appear.

Latin upper case letters
A accidental action; area
B breadth

C factor

E modulus of elasticity; effect of actions; integrity criterion (fire)

F action; force

G permanent action; shear modulus

H horizontal load or force per frame per storey

I second moment of area; thermal insulation criterion (fire)

K stiffness factor (I/L); coefficient

L length; span; system length

M moment in general; bending moment

N axial force

P shear resistance of a shear connector

Q variable action

R resistance; response factor

S stiffness; width (of floor)

V shear force; vertical load per frame per storey

W section modulus; wind load

X property of a material

Z shape factor

Greek upper case letters

Δ difference in . . . (precedes main symbol)

Ψ combination factor for variable action

Latin lower case letters

a dimension; geometrical data; acceleration

b width; breadth; dimension

c outstand; thickness of concrete cover; dimension

d diameter; depth; effective depth

e eccentricity; dimension

f strength (of a material); natural frequency; coefficient; factor

g permanent action; gravitational acceleration

h depth of member

i radius of gyration

k coefficient; factor; property of composite slab; stiffness

l length; buckling length

m property of composite slab; mass per unit length; number

n modular ratio; number

q variable action

r radius; ratio

s spacing

t thickness

u dimension; perimeter

v shear stress; shear force per unit length

w crack width; load per unit length

x dimension to neutral axis; depth of stress block; co-ordinate along member

y major axis; co-ordinate

$\bar{y}$ distance of excluded area from centre of area

z lever arm; dimension; co-ordinate

Greek lower case letters

α angle; ratio; factor

β angle; ratio; factor; coefficient

γ partial factor

δ steel contribution ratio; deflection

ε strain; coefficient

ζ critical damping ratio

η coefficient; degree of shear connection; resistance ratio (fire)

θ angle; slope; temperature

λ (or $\bar{\lambda}$ if non-dimensional) slenderness ratio

μ coefficient of friction; ratio of bending moments; exponent (superscript)

υ Poisson's ratio

ρ reinforcement ratio; density (unit mass)

σ normal stress

τ shear stress

ϕ diameter of a reinforcing bar; rotation; angle of sidesway

χ reduction factor (for buckling)

φ creep coefficient

Subscripts

A   accidental; area; structural steel

a   structural steel; spacing

b   buckling; bolt; beam; bearing

C   concrete

c   compression; concrete; composite

cr   critical

cu   concrete cube

d   design; dynamic

E   effect of action

eff   effective

e   effective (with further subscript); elastic

el   elastic

F   action

f   flange; full shear connection; front; finish (in h_f); full interaction

fi	fire
G	permanent (referring to actions)
g	centroid
H	horizontal
h	haunch
hog	hogging bending
i	index (replacing a numeral); thermal insulation
ini	initial
j	joint
k	characteristic
L	longitudinal (in v_L, shear flow)
LT	lateral-torsional
l	(or ℓ) longitudinal; lightweight-aggregate
M	material; bending moment
m	(allowing for) bending moment; mean; mass
max	maximum
min	minimum
N	(allowing for) axial force
n	number; neutral axis
o	particular value
p	profiled steel sheeting; point (concentrated) load; perimeter; plastic
pa,pr	properties of profiled sheeting (Section 3.3.1)
pl	plastic
Q	variable action
R	resistance
r	reduced; rib
rms	root mean square
S	reinforcing steel
s	reinforcing steel; shear span; slab
sag	sagging bending
sc	shear connector
T	tensile force
t	tension; torsion; time; transverse; top; total
u	ultimate
V	shear
Vs	shear (composite slab)
v	vertical; shear; shear connection
w	web
x	axis along member
y	major axis of cross-section; yield
z	minor axis of cross-section

0, 1, 2, etc.	particular values
0	combination value (in Ψ_0); fundamental (in f_0)
1	frequent value (in Ψ_1); uncracked
2	quasi-permanent value (in Ψ_2); cracked reinforced
0.05, 0.95	fractiles

Chapter 1
Introduction

1.1 Composite beams and slabs

The design of structures for buildings and bridges is mainly concerned with the provision and support of load-bearing horizontal surfaces. Except in some long-span structures, these floors or decks are usually made of reinforced concrete, for no other material provides a better combination of low cost, high strength, and resistance to corrosion, abrasion and fire.

The economical span for a uniform reinforced concrete slab is little more than that at which its thickness becomes sufficient to resist the point loads to which it may be subjected or, in buildings, to provide the sound insulation required. For spans of more than a few metres, it is cheaper to support the slab on beams, ribs or walls than to thicken it. Where the beams or ribs are also of concrete, the monolithic nature of the construction makes it possible for a substantial breadth of slab to act as the top flange of the beam that supports it.

At spans of more than about 10 m, and especially where the susceptibility of steel to loss of strength from fire is not a problem, as in most bridges, steel beams often become cheaper than concrete beams. It was at first customary to design the steelwork to carry the whole weight of the concrete slab and its loading; but by about 1950 the development of shear connectors had made it practicable to connect the slab to the beam, and so to obtain the T-beam action that had long been used in concrete construction. The term 'composite beam' as used in this book refers to this type of structure.

The same term is in use for beams in which prestressed and *in situ* concrete act together; and there are many other examples of composite action in structures, such as between brick walls and beams supporting them, or between a steel-framed shed and its cladding; but these are outside the scope of this book.

No income is received from money invested in construction of a multi-storey building such as a large office block until the building is occupied.

The construction time is strongly influenced by the time taken to construct a typical floor of the building, and here structural steel has an advantage over *in situ* concrete.

Even more time can be saved if the floor slabs are cast on permanent steel formwork, which acts first as a working platform and then as bottom reinforcement for the slab. The use of this formwork, known as *profiled steel sheeting*, commenced in North America [1], and is now standard practice in Europe and elsewhere. These floors span in one direction only, and are known as *composite slabs*. Where the steel sheet is flat, so that two-way spanning occurs, the structure is known as a *composite plate*. These occur in box-girder bridges.

Steel profiled sheeting and partial-thickness precast concrete slabs are known as *structurally participating* formwork. Cement or plastic profiled sheeting reinforced by fibres is sometimes used. Its contribution to the strength of the finished slab is normally ignored in design.

The degree of fire protection that must be provided is another factor that influences the choice between concrete, composite and steel structures, and here concrete has an advantage. Little or no fire protection is required for open multi-storey car parks, a moderate amount for office blocks, and most of all for public buildings and warehouses. Many methods have been developed for providing steelwork with fire protection.

Design against fire and the prediction of fire resistance is known as fire engineering [2]. Several of the Eurocodes have a Part 1.2 devoted to it. Full or partial encasement in concrete is an economical method for steel columns, since the casing makes the columns much stronger. Full encasement of steel beams, once common, is now more expensive than the use of lightweight non-structural materials. Concrete encasement of the web only, done before the beam is erected, is more common in continental Europe than in the UK, and is covered in EN 1994-1-1 [3]. It enhances the buckling resistance of the member (Section 4.2.4) as well as providing fire protection.

The choice between steel, concrete and composite construction for a particular structure thus depends on many factors that are outside the scope of this book. Composite construction is particularly competitive for medium- or long-span structures where a concrete slab or deck is needed for other reasons, where there is a premium for rapid construction, and where a low or medium level of fire protection to steelwork is sufficient.

1.2 Composite columns and frames

When the columns in steel frames were first encased in concrete to protect them from fire, they were designed for the applied load as if uncased. It was then found that encasement reduced the effective slenderness of the column, and so increased its buckling load. Empirical methods for calculating the reduced slenderness still survive in some design codes for steelwork.

This simple approach is not rational, for the encasement also carries its share of both the axial load and the bending moments. More rational methods, validated by tests, are given in EN 1994 (Section 5.6).

A composite column can also be constructed without the use of formwork, by filling a steel tube with concrete. A notable early use of filled tubes (1966) was in a four-level motorway interchange [4]. Their design is covered in Section 5.6.7.

In framed structures, there may be steel members, composite beams, composite columns, or all of these, and there are many types of beam-to-column connection. Their behaviour can range from 'nominally pinned' to 'rigid', and influences bending moments throughout the frame. Two buildings with rigid-jointed composite frames were built in England in the early 1960s, one in Cambridge [5] and one in London [6]. Current practice is mainly to use nominally pinned joints. In buildings it is expensive to make joints so stiff that they act as 'rigid'. Joints are usually treated as pins, even though many have sufficient stiffness to reduce deflections of beams to a useful extent. Intensive research in recent years [7, 8, 9] has enabled comprehensive design rules for joints in steel and composite frames to be given in Eurocodes 3 [10] and 4. Some of them lead to extensive calculation, but they provide the basis for design aids that, when available, may bring semi-rigid connections into general use.

1.3   Design philosophy and the Eurocodes

1.3.1   *Background*

In design, account must be taken of the random nature of loading, the variability of materials, and the defects that occur in construction, to reduce the probability of unserviceability or failure of the structure during its design life to an acceptably low level. Extensive study of this subject since about 1950 has led to the incorporation of the older 'safety factor' and 'load factor' design methods into a comprehensive 'limit state' design philosophy. Its first important application in British standards was in 1972, in CP 110, *The structural use of concrete*. It is used in all current British codes for the design of structures.

Work on international codes began after the Second World War, first on concrete structures and then on steel structures. A committee for composite structures, set up in 1971, prepared the Model Code of 1981 [11]. The Commission of the European Communities has supported work on Eurocodes since 1982, and has delegated its management to the Comité Europeén Normalisation (CEN), based in Brussels. This is an association of the national standards institutions (NSIs) of the countries of the European Union, the European Free Trade Area, and a growing number of other countries from central and eastern Europe.

The Eurocodes EN 1990 to 1999, with over 50 Parts, each with a national annex, are being published by the NSIs, from 2002 until about 2007, as explained in the Preface. Those most relevant to this book are listed as References 3, 10 and 12–16, with the expected or actual date of publication in English by the British Standards Institution. They provide a coherent system, in which duplication of information has been minimised. For example, EN 1994 refers to EN 1990, *Basis of structural design* [12], for design philosophy, most definitions, limit state requirements, and values of partial factors for loads and other actions.

Values for loads and other actions that do not depend on the material used for the structure (the great majority) are given in EN 1991, *Actions on structures* [13]. All provisions for structural steel that apply to both steel and composite structures are in EN 1993, *Design of steel structures* [15]. Similarly, for concrete, EN 1994 refers to but does not repeat material from EN 1992, *Design of concrete structures* [14].

Even within Eurocode 4, material is divided between that which applies to both buildings and bridges, to buildings only, and to bridges only. The first is found in the 'General' clauses of EN 1994-1-1, the second in clauses in EN 1994-1-1 marked 'for buildings', and the third in EN 1994-2, 'Rules for bridges'. Structural fire design is found in EN 1994-1-2 [16], which cross-refers for the high-temperature properties of materials to the 'Fire' parts of EN 1992 and EN 1993, as appropriate.

Design of foundations is covered in EN 1997, *Geotechnical design*, and seismic design in EN 1998, *Design of structures for earthquake resistance*.

This book presents the theories, methods, and models of the 'General' and 'for buildings' rules of Eurocode 4, including relevant material from Eurocodes 1, 2 and 3, but does not refer to or comment on specific clauses. Commentary on EN 1994-1-1 will be found in Reference [17], and on the other codes in the relevant 'Designers' Guides', such as Reference [18].

The British codes current in 2004 that are most relevant to this book are Part 3: Section 3.1 and Part 4 of BS 5950 [19]. They have much in common with EN 1994 as they were developed in parallel with it, but their scope is narrower. For example, columns, web-encased beams, and box girders are not covered.

1.3.2   *Limit state design philosophy*

1.3.2.1   *Basis of design, and actions*

Parts 1.1 of ENs 1992, 1993 and 1994 each have a Section 2, 'Basis of design', that refers to EN 1990 for the presentation of limit state design as used in the Eurocodes. Its Section 4, 'Basic variables', classifies these as actions, environmental influences, properties of materials and products, and geometrical data (e.g., initial out-of-plumb of a column). Actions are either:

- direct actions (forces or loads applied to the structure) or
- indirect actions, such as deformations imposed on the structure, for example by settlement of foundations, change of temperature, or shrinkage of concrete.

'Actions' thus has a wider meaning than 'loads'. Similarly, the Eurocode term 'effect of actions' has a wider meaning than 'stress resultant', because it includes stresses, strains, deformations, crack widths, etc., as well as bending moments, shear forces, etc. The Eurocode term for 'stress resultant' is 'internal force or moment'.

The scope of the following introduction to limit state design is limited to that of the design examples in this book. There are two classes of *limit states*:

- ultimate (denoted ULS), which are associated with structural failure, whether by rupture, crushing, buckling, fatigue or overturning, and
- serviceability (SLS), such as excessive deformation, vibration, or width of cracks in concrete.

Either type of limit state may be reached as a consequence of poor design, construction, or maintenance, or from overloading, insufficient durability, fire, etc.

There are three types of *design situation*:

- persistent, corresponding to normal use;
- transient, for example during construction, refurbishment or repair;
- accidental, such as fire, explosion or earthquake.

There are three main types of action:

- permanent (G or g), such as self-weight of a structure (formerly 'dead load') and including shrinkage of concrete;

- variable (Q or q), such as imposed, wind or snow load (formerly 'live load') and including expected changes of temperature;
- accidental (A), such as impact from a vehicle and high temperature from a fire.

The spatial variation of an action is either:

- fixed (typical of permanent actions) or
- free (typical of other actions), and meaning that the action may occur over only a part of the area or length concerned.

Permanent actions are represented (and specified) by a *characteristic value*, G_k. 'Characteristic' implies a defined fractile of an assumed statistical distribution of the action, modelled as a random variable. For permanent loads, it is usually the mean value (50% fractile).

Variable actions have four *representative values*:

- characteristic (Q_k), normally the upper 5% fractile;
- combination ($\psi_0 Q_k$), for use where the action is assumed to accompany the design ultimate value of another variable action, which is the 'leading action';
- frequent ($\psi_1 Q_k$), for example, occurring at least once a week, and
- quasi-permanent ($\psi_2 Q_k$).

Recommended values for the combination factors ψ_0, ψ_1 and ψ_2 (all less than 1.0) are given in EN 1990. Definitive values, usually those recommended, are given in national annexes. For example, for imposed loads on the floors of offices, the recommended values are $\psi_0 = 0.7$, $\psi_1 = 0.5$, and $\psi_2 = 0.3$.

Design values of actions are, in general, $F_d = \gamma_F F_k$, and in particular,

$$G_d = \gamma_G G_k \tag{1.1}$$

$$Q_d = \gamma_Q Q_k \quad \text{or} \quad Q_d = \gamma_Q \psi_i Q_k \tag{1.2}$$

where γ_G and γ_Q are partial factors for actions, recommended in EN 1990 and given in national annexes. They depend on the limit state considered, and on whether the action is unfavourable or favourable for (i.e., tends to increase or decrease) the action effect considered. The values used in this book are given in Table 1.1.

The *effects of actions* are the responses of the structure to the actions:

$$E_d = E(F_d) \tag{1.3}$$

Table 1.1   Values of γ_G and γ_Q for persistent design situations

Type of action	Permanent unfavourable	Permanent favourable	Variable unfavourable	Variable favourable
Ultimate limit states	1.35*	1.35*	1.5	0
Serviceability limit states	1.0	1.0	1.0	0

*Except for checking loss of equilibrium, or where the coefficient of variation is large

where the function E represents the process of structural analysis. Where the effect is an internal force or moment, *verification for an ultimate limit state* consists of checking that

$$E_d \leq R_d \tag{1.4}$$

where R_d is the relevant design resistance of the system or member or cross-section considered.

1.3.2.2   *Resistances*

Resistances, R_d, are calculated using design values of properties of materials, X_d, given by

$$X_d = X_k/\gamma_M \tag{1.5}$$

where X_k is a characteristic value of the property and γ_M is the partial factor for that property.

The characteristic value is typically a 5% lower fractile (e.g., for compressive strength of concrete). Where the statistical distribution is not well established, it is replaced by a *nominal* value (e.g., the yield strength of structural steel), so chosen that it can be used in design in place of X_k.

The subscript M in γ_M is often replaced by a letter that indicates the material concerned, as shown in Table 1.2, which gives the values of γ_M

Table 1.2   Recommended values for γ_M for strengths of materials and for resistances

Material	Structural steel	Profiled sheeting	Reinforcing steel	Concrete	Shear connection
Property	f_y	f_y	f_{sk}	f_{ck} or f_{cu}	P_{Rk}
Symbol for γ_M	γ_A	γ_A	γ_S	γ_C	γ_V or γ_{Vs}
Ultimate limit states	1.0	1.0	1.15	1.5	1.25
Serviceability limit states	1.0	1.0	1.0	1.0	1.0

Notation: for concrete, f_{ck} and f_{cu} are respectively characteristic cylinder and cube strengths; symbol γ_{Vs} is for shear resistance of a composite slab.

used in this book. A welded stud shear connector is treated like a single material, even though its design resistance to shear, P_{Rk}/γ_V, is influenced by the properties of both steel and concrete. For resistance to fracture of a steel cross-section in tension, $\gamma_A = 1.25$.

1.3.2.3 *Combinations of actions*

The Eurocodes treat systematically a subject for which many empirical procedures have been used in the past. For ultimate limit states, the principles are:

- permanent actions are present in all combinations;
- each variable action is chosen in turn to be the 'leading' action (i.e., to have its full design value) and is combined with lower 'combination' values of other variable actions that may co-exist with it;
- the design action effect is the most unfavourable of those found by this process.

The use of combination values allows for the limited correlation between time-dependent variable actions.

As an example, it is assumed that a bending moment M_{Ed} in a member is influenced by its own weight, G, by an imposed vertical load, Q_1, and by wind loading, Q_2. The fundamental combinations for verification for persistent design situations are:

$$\gamma_G G_k + \gamma_{Q1} Q_{k,1} + \gamma_{Q2} \psi_{0,2} Q_{k,2} \tag{1.6}$$

and

$$\gamma_G G_k + \gamma_{Q1} \psi_{0,1} Q_{k,1} + \gamma_{Q2} Q_{k,2} \tag{1.7}$$

Each term in these expressions gives the value of the action for which a bending moment is calculated, and the + symbols apply to the bending moments, not to the values of the actions. This is sometimes indicated by placing each term between quotation marks.

In practice, it is usually obvious which combination will govern. For low-rise buildings, wind is rarely critical for floors, so Expression 1.6 with imposed load leading would be used; but for a long-span lightweight roof, Expression 1.7 would govern, and both positive and negative wind pressure would be considered, with the negative pressure combined with $Q_{k,1} = 0$.

The combination for the accidental design situation of 'fire' is given in Section 3.3.7.

Table 1.3 Combination factors

Factor	ψ_0	ψ_1	ψ_2
Imposed floor loading in an office area of a building	0.7	0.5	0.3
Wind loading on a building	0.6	0.2	0

For serviceability limit states, three combinations are defined. The most onerous, the 'characteristic' combination, corresponds to the fundamental combination (above) with the γ factors reduced to 1.0. For the example in Expressions 1.6 and 1.7, it is

$$G_k + Q_{k,1} + \psi_{0,2}Q_{k,2} \tag{1.8}$$

and

$$G_k + \psi_{0,1}Q_{k,1} + Q_{k,2} \tag{1.9}$$

It is normally used for verifying irreversible limit states, for example, deformations that result from the yielding of steel.

Assuming that Q_1 is the leading variable action, the others are:

- frequent combination,

$$G_k + \psi_{1,1}Q_{k,1} + \psi_{2,2}Q_{k,2} \tag{1.10}$$

- quasi-permanent combination,

$$G_k + \psi_{2,1}Q_{k,1} + \psi_{2,2}Q_{k,2} \tag{1.11}$$

The frequent combination is used for reversible limit states, for example, the elastic deflection of a floor under imposed loading. However, if that deformation causes cracking of a brittle floor finish or damage to fragile partitions, then the limit state is not reversible, and the check should be done for the higher (less probable) loading of the characteristic combination.

The quasi-permanent combination is used for long-term effects (e.g., deformations from creep of concrete) and for the appearance of the structure.

Some combination factors used in this book are given in Table 1.3.

1.3.2.4 *Comments on limit state design philosophy*

The use of limit states has superseded earlier methods, partly because limit states provide identifiable criteria for satisfactory performance.

Stresses cannot be calculated with the same confidence as resistances of members, and high values may or may not be significant.

An apparent disadvantage of limit states design is that several sets of design calculations may be needed whereas, with some older methods, one was sufficient. This is only partly true, for it has been found possible to identify many situations in which design for, say, an ultimate limit state will ensure that certain types of unserviceability will not occur, and vice versa. In the rules of EN 1994 for buildings it has generally been possible to avoid specifying limiting stresses for serviceability limit states by using other methods to check deflections and crack widths.

1.4   Properties of materials

Information on the properties of structural steel, profiled sheeting, concrete and reinforcement is readily available. Only that which has particular relevance to composite structures is given here.

In the determination of the bending moments and shear forces in a beam or framed structure (known as 'global analysis'), all the materials can be assumed to behave in a linear-elastic manner, though an effective modulus is used for the concrete to allow for its creep under sustained compressive stress. Its tensile strength need not be taken as zero, provided account is taken of reductions of stiffness caused by cracking. The effects of its shrinkage are rarely significant in buildings.

Rigid-plastic global analysis can sometimes be used (Section 4.3.3) despite the profound difference between a typical stress–strain curve for concrete in compression and those for structural steel or reinforcement, in tension or compression, that is illustrated in Fig. 1.1.

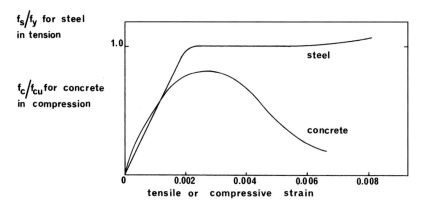

Figure 1.1   Stress–strain curves for concrete and structural steel

Concrete reaches its maximum compressive stress at a strain of between 0.002 and 0.003, and at higher strains it crushes, losing almost all of its compressive strength. It is very brittle in tension, having a strain capacity of only about 0.0001 (i.e., 0.1 mm per metre) before it cracks. Figure 1.1 also shows that the maximum stress reached by concrete in a beam or column is well below its cube strength.

Steel yields at a strain similar to that given for the maximum stress in concrete, but on further straining the stress continues to increase slowly, until (for a typical structural steel) the total strain is at least thirty times the yield strain. Its subsequent necking and fracture is of little significance for composite members because the useful resistance of a cross-section is reached when all of the steel has yielded, when steel in compression buckles, or when concrete crushes.

Resistances of cross-sections are determined using plastic analysis wherever possible because results of elastic analyses are unreliable, unless careful account is taken of the cracking, shrinkage, and creep of concrete (which is difficult), and also because plastic analysis is simpler and leads to more economical design.

The use of a higher value of γ_M for concrete than for steel (Table 1.2) includes allowance for the higher variability of the strength of test specimens, and the variation in the strength of concrete over the depth of a member, due to migration of water before setting. It also allows for the larger errors in the dimensions of cross-sections, particularly in the positions of reinforcing bars.

Brief comments are now given on individual materials.

Concrete

In EN 1992, a typical strength class for normal-density concrete is denoted C25/30, where the specified characteristic compressive strengths at age 28 days are $f_{ck} = 25$ N/mm^2 (cylinder test) and $f_{cu} = 30$ N/mm^2 (cube test). The design formulae in EN 1994 use f_{cd}, which is f_{ck}/γ_C. The normal-density concrete used in worked examples here is 'Grade 30' (in British terminology), with f_{ck} taken as 25 N/mm^2. It is used for the columns and for encasement of beam webs. The floor slabs are constructed with lightweight-aggregate concrete of grade LC25/28, with oven-dry density 1800 kg/m^3. Other properties for these two concretes are given in Table 1.4.

For densities, EN 1991-1-1 uses kN/m^3 units, and so gives densities about 2% higher than those of EN 1992-1-1, since 1800 kg/m^3 is 17.65 kN/m^3. The higher values are used here.

Reinforcing steel

Strength grades for reinforcing steel are given in EN 10080 [20] in terms of a characteristic yield strength f_{sk}. The value used here is 500 N/mm^2,

Table 1.4 Properties of concretes used in the examples, at age 28 days

Concrete grade			C25/30	LC25/28
Characteristic cylinder strength,	N/mm^2	f_{ck}	25	25
Mean cylinder strength,	N/mm^2	f_{cm}	33	33
Lower tensile strength,	N/mm^2	$f_{ct,0.05}$	1.80	1.60
Mean tensile strength,	N/mm^2	f_{ctm}	2.60	2.32
Upper tensile strength,	N/mm^2	$f_{ct,0.95}$	3.30	2.94
Mean elastic modulus,	kN/mm^2	E_{cm}	31.0	20.7
Weight density, reinforced concrete, kN/m^3			25.0	19.5

for both ribbed bars and welded steel fabric. The modulus of elasticity for reinforcement, E_s, is normally taken as 200 kN/mm^2; but in a composite section it may be assumed to have the value for structural steel, E_a = 210 kN/mm^2, as the error is negligible.

Structural steel

Strength grades for structural steel are given in EN 10025 [21] in terms of a nominal yield strength, f_y, and ultimate tensile strength, f_u. The grade used in worked examples here is S355, for which f_y = 355 N/mm^2, f_u = 510 N/mm^2 for elements of all thicknesses up to 40 mm.

The density of structural steel is assumed to be 7850 kg/m^3. Its coefficient of linear thermal expansion is 12×10^{-6} per °C. The difference between this value and that for normal-density concrete, 10×10^{-6} per °C, can usually be ignored.

Profiled steel sheeting

This product is available with yield strengths ranging from 235 N/mm^2 to at least 460 N/mm^2, in profiles with depths ranging from 45 mm to over 200 mm, and with a wide range of shapes. These include both re-entrant ribs, and trapezoidal troughs as in Fig. 3.9. There are various methods of achieving composite action with a concrete slab, discussed in Section 2.4.3.

Sheets are normally between 0.8 mm and 1.5 mm thick, and are protected from corrosion by a zinc coating about 0.02 mm thick on each face. Elastic properties of the material may be assumed to be as for structural steel.

Shear connectors

In the early years of composite construction, many types of connector were in use. This market is now dominated by automatically-welded headed studs. Details of these and the measurement of their resistance to shear are given in Chapter 2.

1.5 Direct actions (loading)

The characteristic loadings to be used in worked examples are now given. They are taken from EN 1991.

The *permanent loads* (dead load) are the weight of the structure and its finishes. The structural steel or profiled sheeting component of a composite member is invariably built first, so a distinction must be made between load resisted by the steel component only, and load applied to the member after the concrete has developed sufficient strength for composite action to become effective. The division of the permanent load between these categories depends on the method of construction.

Composite beams and slabs are classified as *propped* ('shored' in North America) or *unpropped*. In propped construction, the steel member is supported at intervals along its length until the concrete has reached a certain proportion, usually three-quarters, of its design strength. When the props are removed, the whole of the dead load is assumed to be carried by the composite member. Where no props are used, it is assumed in elastic analysis that the steel member alone carries its own weight and that of the formwork and the concrete slab. Other dead loads such as floor finishes and internal walls are added later, and so are assumed to be carried by the composite member. In ultimate-strength methods of analysis (Section 3.5.3), inelastic behaviour causes extensive redistribution of stress before failure, and it can be assumed that the whole load is applied to the composite member, whatever the method of construction.

The principal vertical *variable action* in a building is a uniformly-distributed load on each floor. EN 1991-1-1 gives ranges of loads, depending on the use to be made of the area, with a recommended value. For 'office areas', this value is

$$q_k = 3.0 \text{ kN/m}^2 \qquad (1.12)$$

Account is taken of point loads (e.g., a safe being moved on a trolley with small wheels) by defining an alternative point load, to be applied anywhere on the floor, on an area about 50 mm square. For the type of area defined above, this is

$$Q_k = 4.5 \text{ kN} \qquad (1.13)$$

Where a member such as a column is carrying loads q_k from n storeys ($n > 2$), the total of these loads may be multiplied by a reduction factor α_n. The recommended value is

$$\alpha_n = [2 + (n - 2)\psi_0]/n \qquad (1.14)$$

where ψ_0 is the combination factor (e.g., as in Table 1.3). This allows for the low probability that all n floors will be fully loaded at once.

In an office block, the location of partitions is unknown at the design stage. Their weight is usually allowed for by increasing the imposed loading, q_k, by an amount that depends on the expected weight per unit length of the partitions. The increases given in EN 1991-1-1 range from 0.5 to 1.2 kN/m².

The principal horizontal variable load for a building is wind. These loads are given in EN 1991-1-4. They usually consist of pressure or suction on each external surface. On large flat areas, frictional drag may also be significant. Wind loads rarely influence the design of composite beams, but can be important in framed structures not braced against side-sway and in tall buildings.

Methods of calculation for distributed and point loads are sufficient for all types of direct action. Indirect actions such as subsidence or differential changes of temperature, which occasionally influence the design of structures for buildings, are not considered in this book.

1.6   Methods of analysis and design

The purpose of this section is to provide a preview of the principal methods of analysis used for composite members and frames, and to show that most of them are straightforward applications of methods in common use for steel or reinforced concrete structures.

The steel designer will be familiar with the elementary elastic theory of bending, and the simple plastic theory in which the whole cross-section of a member is assumed to be at yield, in either tension or compression. Both theories are used for composite members, the differences being as follows:

- concrete in tension is usually neglected in elastic theory, and always neglected in plastic theory;
- in the elastic theory, concrete in compression is 'transformed' into an equivalent area of steel by dividing its breadth by the modular ratio E_a/E_c;
- in the plastic theory, the design 'yield stress' of concrete in compression is taken as $0.85f_{cd}$, where $f_{cd} = f_{ck}/\gamma_C$. Transformed sections are not used. Examples of this method are given in Sections 3.4.2 and 3.11.1.

In the strength classes for concrete in EN 1992, the ratios f_{ck}/f_{cu} range from 0.78 to 0.83, so for $\gamma_C = 1.5$, the stress $0.85f_{cd}$ corresponds to a value between $0.44f_{cu}$ and $0.47f_{cu}$. This agrees closely with BS 5950 [19], which uses $0.45f_{cu}$ for the plastic resistance of cross-sections.

The factor 0.85 takes account of several differences between a standard cylinder test and what concrete experiences in a structural member. These include the longer duration of loading in the structure, the presence of a stress gradient across the section considered, and the differences in the boundary conditions for the concrete.

The concrete designer will be familiar with the method of transformed sections, and with the rectangular-stress-block theory outlined above. The basic difference from the elastic behaviour of a reinforced concrete beam is that the steel section in a composite beam is more than tension re-inforcement. It has a significant bending stiffness of its own, and resists most of the vertical shear.

The formulae for the elastic properties of composite sections are more complex than those for steel or reinforced concrete sections. The chief reason is that the neutral axis for bending may lie in the web, the steel flange, or the concrete flange of the member. The theory is not in prin-ciple any more complex than that for a steel I-beam.

Longitudinal shear

Students usually find this subject troublesome, even though the formula

$$\tau = VA\bar{y}/Ib \tag{1.15}$$

is familiar from their study of vertical shear in elastic beams, so a note on its use may be helpful. Its proof can be found in any undergraduate-level textbook on strength of materials.

First, let us consider the shear stresses τ in the elastic I-beam shown in Fig. 1.2, due to a vertical shear force V. For a cross-section 1–2 through the web, the 'excluded area', A in the formula, is the top flange, of area A_f. The distance $\bar{y}$ of its centroid from the neutral axis (line X–X in Fig. 1.2) is $(h - t_f)/2$. The shear stress τ_{12} on plane 1–2, of breadth t_w, is therefore

$$\tau_{12} = VA_f(h - t_f)/(2It_w) \tag{1.16}$$

where I is the second moment of area of the whole cross-section about the axis X–X through its centre of area.

This result may be recognised as the *vertical* shear stress at this cross-section. However, a shear stress is always associated with a complemen-tary shear stress at right angles to it and of equal value; in this case, the *longitudinal* shear stress. This will now be denoted by v, rather than τ, as v is used in EN 1994.

If the cross-section in Fig. 1.2 is a composite beam, with the cross-hatched area representing the transformed area of a concrete flange, shear connection is required on plane 1–2. It has to resist this longitudinal shear

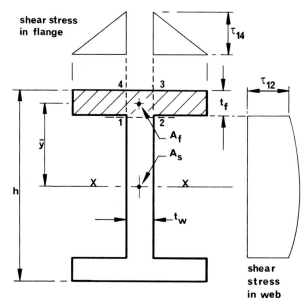

Figure 1.2   Shear stresses in an elastic I-section

stress over a width t_w, so the shear force per unit length is vt_w. This is named the *shear flow*, and has the symbol v_L. From these results,

$$v_{L,12} = VA_f(h - t_f)/(2I) \tag{1.17}$$

Consideration of the longitudinal equilibrium of the small element 1234 in Fig. 1.2 shows that if its area $t_w t_f$ is much less than A_f, the shear flows on planes 1–4 and 2–3 are each approximately $v_{L,12}/2$, and the mean shear stress on these planes is given approximately by

$$\tau_{14} t_f = \tau_{12} t_w/2$$

This stress is needed for checking the resistance of the concrete slab to longitudinal shear.

Repeated use of Equation 1.15 for various cross-sections shows that the variation of longitudinal shear stress is parabolic in the web and linear in the flanges, as shown in Fig. 1.2.

The second example is the elastic beam shown in section in Fig. 1.3. This represents a composite beam in sagging bending, with the neutral axis at depth x, a concrete slab of thickness h_c, and the interface between the slab and the structural steel (which is assumed to have no top flange) at level 6–5. The concrete has been transformed to steel, so the cross-hatched area is the equivalent steel section. The concrete in area ABCD is

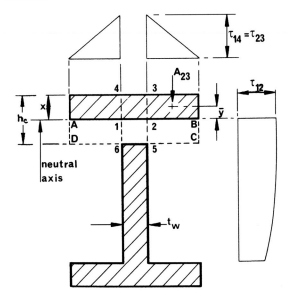

Figure 1.3 Shear stresses in a composite section with the neutral axis in the concrete slab

assumed to be cracked. As in the theory for reinforced concrete beams, it resists no longitudinal stress but is capable of transferring shear stress.

Equation 1.15 is based on rate of change along the beam of bending stress, so in applying it here, area ABCD is omitted when the 'excluded area' is calculated. Let the cross-hatched area of flange be A_f, as before. The longitudinal shear stress on plane 6–5 is given by

$$v_{65} = VA_f\bar{y}/It_w \tag{1.18}$$

where $\bar{y}$ is the distance from the centroid of the excluded area to the neutral axis, *not to plane 6–5*. If A and $\bar{y}$ are calculated for the cross-hatched area below plane 6–5, the same value v_{65} is obtained, because it is the equality of the two products '$A\bar{y}$' that determines the value x.

The preceding theory relies on the assumption that the flexibility of shear connectors is negligible, and is used in bridge design and for fatigue generally. Ultimate-strength theory (Sections 3.3.2 and 3.6.2) provides an alternative that takes advantage of the plastic behaviour of stud connectors and is widely used in design for buildings.

For a plane such as 2–3 in Fig. 1.3, the shear flow is

$$v_{L,23} = VA_{23}\bar{y}/I \tag{1.19}$$

The design shear stress for the concrete on this plane is $v_{L,23}/h_c$. It is not equal to $v_{L,23}/x$ because the cracked concrete can resist shear. The depth h_c

does not have to be divided by the modular ratio, even though the transformed section is of steel, because the transformation is of widths, not depths. An alternative explanation is that shear flows from equations such as Equation 1.19 are independent of the material considered, because transformation does not alter the ratio A_{23}/I.

Longitudinal slip

Shear connectors are not rigid, so that a small longitudinal slip occurs between the steel and concrete components of a composite beam. The problem does not arise in other types of structure, and relevant analyses are quite complex (Section 2.6 and Appendix A). They are not needed in design, for which simplified methods have been developed.

Deflections

The effects of creep and shrinkage make calculation of deflections in reinforced concrete beams more complex than for steel beams, but the limiting span/depth ratios given in codes such as BS 8110 [22] provide a simple means of checking for excessive deflection. These ratios are unreliable for composite beams, especially where unpropped construction is used. Examples of checks on deflections are given in Sections 3.4.5 and 3.11.3.

Vertical shear

The stiffness in vertical shear of the concrete slab of a composite beam is usually much less than that of the steel component, and is neglected in design. For vertical shear, the methods used for steel beams are applicable also to composite beams.

Buckling of flanges and webs of beams

This will be a new problem to many designers of reinforced concrete. It leads to restrictions on the breadth/thickness ratios of unstiffened steel webs and flanges in compression (Section 3.5.2). These do not apply to the steel part of the top flange of a composite T-beam at mid-span, because local buckling is prevented by its attachment to the concrete slab.

Crack-width control

The maximum spacings for reinforcing bars given in codes for reinforced concrete are intended to limit the widths of cracks in the concrete, for reasons of appearance and to avoid corrosion of reinforcement. In composite structures for buildings, cracking is likely to be a problem only where the top surfaces of continuous beams support brittle finishes or are exposed to corrosion. The principles of crack-width control are the same as for reinforced concrete. The calculations are different, but can normally be avoided by using the simplified methods given in EN 1994.

Continuous beams

The Eurocode design methods for continuous beams (Chapter 4) make use of both simple plastic theory (as for steel beams) and redistribution of moments (as for concrete beams).

Columns

The only British code that gives a design method for composite columns is BS 5400:Part 5, *Composite bridges*. EN 1994 gives a newer and simpler method, which is described in Section 5.6.

Framed structures for buildings

Composite members normally form part of a frame that is essentially steel, rather than concrete, so the design methods of EN 1994 are based on those of EN 1993 for steel structures. Beam-to-column joints are classified in the same way; the same assumptions are made about geometrical imperfections, such as out-of-plumb columns; and similar allowance is made for second-order effects (increases in bending moments and reduction in stability, caused by interaction between vertical loads and lateral deflections). Frame analysis is outlined in Section 5.4.5. It may be more complex than in current practice, but includes methods for unbraced frames. Eurocodes EN 1993 and 1994 contain much new material on the design of joints.

Structural fire design

The high thermal conductivity and slenderness of structural steel members and profiled sheeting cause them to lose strength in fire more quickly than concrete members do. Structures for buildings are required to have fire resistance of minimum duration (typically, 30 minutes to 2 hours) to enable occupants to escape and to protect fire fighters. This has led to the provision of minimum thicknesses of concrete and areas of reinforcement and of thermal insulation of steelwork.

Extensive research and the recent subject of fire engineering [2] have enabled the Eurocode rules for resistance to fire to be less onerous than older rules. Advantage is taken in design of membrane effects associated with large deformations, and of the provisions for accidental design situations. These allow for over-strength of members and the use of 'frequent' rather than 'characteristic' load levels. Explanations and worked examples are given in Sections 3.3.7, 3.4.6, 3.10, 3.11.4 and 5.6.2.

Chapter 2
Shear connection

2.1 Introduction

The established design methods for reinforced concrete and for structural steel give no help with the basic problem of connecting steel to the concrete. The force applied to this connection is mainly, but not entirely, longitudinal shear. As with bolted and welded joints, the connection is a region of severe and complex stress that defies accurate analysis, and so methods of connection have been developed empirically and verified by tests. They are described in Section 2.4.

The simplest type of composite member used in practice occurs in floor structures of the type shown in Fig. 3.1. The concrete floor slab is continuous over the steel I-sections, and is supported by them. It is designed to span in the y-direction in the same way as when supported by walls or the ribs of reinforced concrete T-beams. When shear connection is provided between the steel member and the concrete slab, the two together span in the x-direction as a composite beam. The steel member has not been described as a 'beam', because its main function at mid-span is to resist tension, as does the reinforcement in a T-beam. The compression is assumed to be resisted by an 'effective' width of slab, as explained in Section 3.5.1.

In buildings, but not in bridges, these concrete slabs are often composite with profiled steel sheeting (Fig. 2.8), which rests on the top flange of the steel beam. Other types of cross-section that can occur in composite beams are shown in Fig. 2.1.

The ultimate-strength design methods used for shear connection in beams and columns in buildings are described in Sections 3.6 and 5.6.6, respectively.

The subjects of the present chapter are: the effects of shear connection on the behaviour of very simple beams, current methods of shear connection, standard tests on shear connectors, and shear connection in composite slabs.

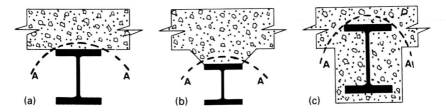

Figure 2.1 Typical cross-sections of composite beams

2.2   Simply-supported beam of rectangular cross-section

Flitched beams, whose strength depended on shear connection between parallel timbers, were used in mediaeval times, and survive today in the form of glued-laminated construction. Such a beam, made from two members of equal size (Fig. 2.2), will now be studied. It carries a load w per unit length over a span L, and its components are made of an elastic material with Young's modulus E. The weight of the beam is neglected.

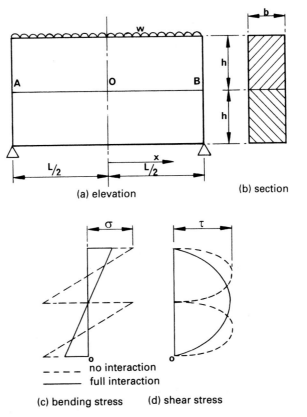

Figure 2.2 Effect of shear connection on bending and shear stresses

2.2.1 *No shear connection*

It is assumed first that there is no shear connection or friction on the interface AB. The upper beam cannot deflect more than the lower one, so each carries load $w/2$ per unit length as if it were an isolated beam of second moment of area $bh^3/12$, and the vertical compressive stress across the interface is $w/2b$. The mid-span bending moment in each beam is $wL^2/16$. By elementary beam theory, the stress distribution at mid-span is given by the dashed line in Fig. 2.2(c), and the maximum bending stress in each component, σ, is given by

$$\sigma = \frac{My_{max}}{I} = \frac{wL^2}{16}\frac{12}{bh^3}\frac{h}{2} = \frac{3wL^2}{8bh^2} \tag{2.1}$$

The maximum shear stress, τ, occurs near a support. The two parabolic distributions given by simple elastic theory are shown in Fig. 2.2(d); and at the centre-line of each member,

$$\tau = \frac{3}{2}\frac{wL}{4}\frac{1}{bh} = \frac{3wL}{8bh} \tag{2.2}$$

The maximum deflection, δ, is given by the usual formula

$$\delta = \frac{5(w/2)L^4}{384EI} = \frac{5}{384}\frac{w}{2}\frac{12L^4}{Ebh^3} = \frac{5wL^4}{64Ebh^3} \tag{2.3}$$

The bending moment in each beam at a section distant x from mid-span is $M_x = w(L^2 - 4x^2)/16$, so that the longitudinal strain ε_x at the bottom fibre of the upper beam is

$$\varepsilon_x = \frac{My_{max}}{EI} = \frac{3w}{8Ebh^2}(L^2 - 4x^2) \tag{2.4}$$

There is an equal and opposite strain in the top fibre of the lower beam, so that the difference between the strains in these adjacent fibres, known as the *slip strain*, is $2\varepsilon_x$.

It is easy to show by experiment with two or more flexible wooden laths or rulers that, under load, the end faces of the two-component beam have the shape shown in Fig. 2.3(a). The slip at the interface, s, is zero at $x = 0$ (from symmetry) and a maximum at $x = \pm L/2$. The cross-section at $x = 0$ is the only one where plane sections remain plane. The slip strain, defined above, is not the same as slip. In the same way that strain is rate of change of displacement, slip strain is the rate of change of slip along the beam. Thus from Equation 2.4,

$$\frac{ds}{dx} = 2\varepsilon_x = \frac{3w}{4Ebh^2}(L^2 - 4x^2) \tag{2.5}$$

Integration gives

$$s = \frac{w}{4Ebh^2}(3L^2x - 4x^3) \tag{2.6}$$

The constant of integration is zero, since $s = 0$ when $x = 0$, so that Equation 2.6 gives the distribution of slip along the beam.

Results (2.5) and (2.6) for the beam studied in Section 2.7 are plotted in Fig. 2.3. This shows that at mid-span, slip strain is a maximum and slip is zero and, at the ends of the beam, slip is a maximum and slip strain is zero. From Equation 2.6, the maximum slip (when $x = L/2$) is $wL^3/4Ebh^2$. Some idea of the magnitude of this slip is given by relating it to the maximum deflection of the two beams. From Equation 2.3, the ratio of

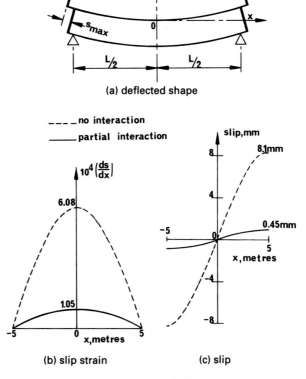

(a) deflected shape

(b) slip strain

(c) slip

Figure 2.3   Deflections, slip strain and slip

slip to deflection is $3.2h/L$. The ratio $L/2h$ for a beam is typically about 20, so that the end slip is less than a tenth of the deflection. This shows that *shear connection must be very stiff if it is to be effective.*

2.2.2 Full interaction

It is now assumed that the two halves of the beam shown in Fig. 2.2 are joined together by an infinitely stiff shear connection. The two members then behave as one. Slip and slip strain are everywhere zero, and it can be assumed that plane sections remain plane. This situation is known as *full interaction*. Except in design with partial shear connection (Sections 3.5.3 and 3.7.1), all design of composite beams and columns in practice is based on the assumption that full interaction is achieved.

For the composite beam of breadth b and depth $2h$, $I = 2bh^3/3$, and elementary theory gives the mid-span bending moment as $wL^2/8$. The extreme fibre bending stress is

$$\sigma = \frac{My_{max}}{I} = \frac{wL^2}{8}\frac{3}{2bh^3}h = \frac{3wL^2}{16bh^2} \tag{2.7}$$

The vertical shear at section x is

$$V_x = wx \tag{2.8}$$

so the shear stress at the neutral axis is

$$\tau_x = \frac{3}{2}wx\frac{1}{2bh} = \frac{3wx}{4bh} \tag{2.9}$$

and the maximum shear stress is

$$\tau = \frac{3wL}{8bh} \tag{2.10}$$

The stresses are compared in Figs 2.2(c) and (d) with those for the non-composite beam. The provision of the shear connection does not change the maximum shear stress, but the maximum bending stress is halved.

The mid-span deflection is

$$\delta = \frac{5wL^4}{384EI} = \frac{5wL^4}{256Ebh^3} \tag{2.11}$$

which is one-quarter of the previous deflection (Equation 2.3). Thus the provision of shear connection increases both the strength and the stiffness of a beam of given size, and in practice leads to a reduction in the size of the beam required for a given loading, and usually to a reduction in its cost.

In this example – but not always – the interface AOB coincides with the neutral axis of the composite member, so that the maximum longitudinal shear stress at the interface is equal to the maximum vertical shear stress, which occurs at $x = \pm L/2$ and is $3wL/8bh$, from Equation 2.10.

The shear connection must be designed for the longitudinal shear per unit length, v_L, which is known as the *shear flow*. In this example it is given by

$$v_{L,x} = \tau_x b = \frac{3wx}{4h} \tag{2.12}$$

The total shear flow in a half span is found, by integration of equation (2.12), to be $3wL^2/(32h)$. Typically, $L/2h \simeq 20$, so the shear connection in the whole span has to resist a total shear force

$$2 \times \frac{3}{32}\frac{L}{h}wl \simeq 8wL$$

Thus, this shear force is eight times the total load carried by the beam. A useful rule of thumb is that the resistance of the shear connection for a beam should be an order of magnitude greater than the load to be carried; it shows that *shear connection must be very strong*.

In elastic design, the shear connectors are spaced in accordance with the shear flow. Thus, if the design shear resistance of a connector is P_{Rd}, the pitch or spacing at which they should be provided, p, is given by $pv_{L,x} \not> P_{Rd}$. From Equation 2.12 this is

$$p \not> \frac{4P_{Rd}h}{3wx} \tag{2.13}$$

This is known as 'triangular' spacing, from the shape of the graph of v_L against x (Fig. 2.4).

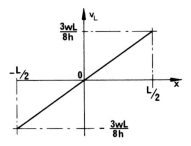

Figure 2.4   Shear flow for 'triangular' spacing of connectors

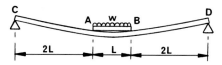

Figure 2.5 Uplift forces

2.3 Uplift

In the preceding example, the stress normal to the interface AOB (Fig. 2.2) was everywhere compressive, and equal to $w/2b$ except at the ends of the beam. The stress would have been tensile if the load w had been applied to the lower member. Such loading is unlikely, except when travelling cranes are suspended from the steelwork of a composite floor above; but there are other situations in which stresses tending to cause uplift can occur at the interface. These arise from complex effects such as the torsional stiffness of reinforced concrete slabs forming flanges of composite beams, the triaxial stresses in the vicinity of shear connectors and, in box-girder bridges, the torsional stiffness of the steel box.

Tension across the interface can also occur in beams of non-uniform section or with partially completed flanges. Two members without shear connection, as shown in Fig. 2.5, provide a simple example. AB is supported on CD and carries distributed loading. It can easily be shown by elastic theory that if the flexural rigidity of AB exceeds about one-tenth of that of CD, then the whole of the load on AB is transferred to CD at points A and B, with separation of the beams between these points. If AB were connected to CD, there would be uplift forces at mid-span.

Almost all connectors used in practice are therefore so shaped that they provide resistance to uplift as well as to slip. Uplift forces are so much less than shear forces that it is not normally necessary to calculate or estimate them for design purposes, provided that connectors with some uplift resistance are used.

2.4 Methods of shear connection

2.4.1 *Bond*

Until the use of deformed bars became common, most of the reinforcement for concrete consisted of smooth mild-steel bars. The transfer of shear from steel to concrete was assumed to occur by bond or adhesion at the concrete–steel interface. Where the steel component of a composite member is surrounded by reinforced concrete, as in an encased beam,

Fig. 2.1(c), or an encased stanchion, Fig. 5.14, the analogy with reinforced concrete suggests that no shear connectors need be provided.
Tests have shown that this can be true for cased stanchions and filled
tubes, where bond stresses are usually low, and also for encased beams
in the elastic range. In design it is necessary to restrict bond stress to a
low value, to provide a margin for the incalculable effects of shrinkage of
concrete, poor adhesion to the underside of steel surfaces, and stresses
due to variations of temperature.

Research on the ultimate strength of encased beams has shown that,
at high loads, calculated bond stresses have little meaning, due to the
development of cracking and local bond failures. If longitudinal shear
failure occurs, it is invariably on a surface such as AA in Fig. 2.1(c), and
not around the perimeter of the steel section. For these reasons, codes of
practice do not allow ultimate-strength design methods to be used for
composite beams without shear connectors.

Most composite beams have cross-sections of types (a) or (b) in Fig. 2.1.
Tests on such beams show that, at low loads, most of the longitudinal shear
is transferred by bond at the interface, that bond breaks down at higher
loads, and that once broken it cannot be restored. So in design calculations, bond strength is taken as zero and, in research, the bond is deliber-
ately destroyed by greasing the steel flange before the concrete is cast. For
uncased beams, the most practicable form of shear connection is some form
of dowel welded to the top flange of the steel member and subsequently
surrounded by *in situ* concrete when the floor or deck slab is cast.

2.4.2 *Shear connectors*

The most widely used type of connector is the headed stud (Fig. 2.6).
These range in diameter from 13 to 25 mm, and in length, h, from 65 to
150 mm, though longer studs are sometimes used. Studs should have an

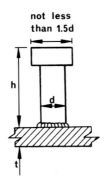

Figure 2.6 Headed stud shear connector

ultimate tensile strength of at least 450 N/mm² and an elongation of at least 15%. The advantages of stud connectors are that the welding process is rapid, they provide little obstruction to reinforcement in the concrete slab, and they are equally strong and stiff in shear in all directions normal to the axis of the stud.

There are two factors that influence the diameter of studs. One is the welding process, which becomes increasingly difficult and expensive at diameters exceeding 20 mm, and the other is the thickness, t (Fig. 2.6), of the plate or flange to which the stud is welded. A study made in the USA [23] found that the full static strength of the stud can be developed if d/t is less than about 2.7, and a limit of 2.5 is given in EN 1994-1-1. Tests using repeated loading have led to the rule that where the flange plate is subjected to fluctuating tensile stress, d/t may not exceed 1.5. These rules prevent the use of welded studs as shear connection in composite slabs.

The maximum shear force that can be resisted by a 25-mm stud is relatively low, about 130 kN. Other types of connector with higher strength have been developed, primarily for use in bridges. These are bars with hoops (Fig. 2.7(a)), tees with hoops, horseshoes and channels (Fig. 2.7(b)).

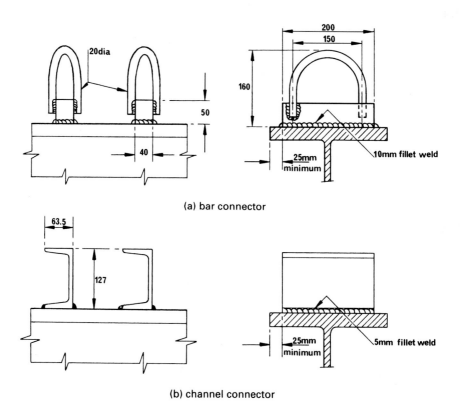

(a) bar connector

(b) channel connector

Figure 2.7   Other types of shear connector

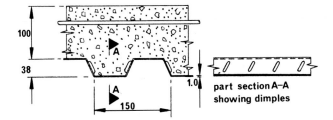

Figure 2.8   Composite slab

Bars with hoops are the strongest of these, with ultimate shear strengths up to 1000 kN. Design rules are given in BS 5400:Part 5 and in the preliminary Eurocode 4, ENV 1994-1-1, but were omitted from EN 1994-1-1 because they are now rarely used. Epoxy adhesives have been tried, but it is not clear how resistance to uplift can reliably be provided where the slab is attached to the steel member only at its lower surface.

2.4.3   *Shear connection for profiled steel sheeting*

This material is commonly used as permanent formwork for floor slabs in buildings, then known as *composite slabs*. Typical cross-sections are shown in Figs 2.8, 2.14, 2.20 and 3.12. As it is impracticable to weld shear connectors to material that may be less than 1 mm thick, shear connection is provided either by pressed or rolled dimples that project into the concrete, or by giving the steel profile a re-entrant shape that prevents separation of the steel from the concrete.

The resistance of composite slabs to longitudinal shear is covered in Section 2.8, and their design in Section 3.3.

2.5   Properties of shear connectors

The property of a shear connector most relevant to design is the relationship between the shear force transmitted, P, and the slip at the interface, s. This load–slip curve should ideally be found from tests on composite beams, but in practice a simpler specimen is necessary. Most of the data on connectors have been obtained from various types of 'push-out' or 'push' test. The flanges of a short length of steel I-section are connected to two small concrete slabs. The details of the standard push test of EN 1994-1-1 are shown in Fig. 2.9. The slabs are bedded onto the lower platen of a compression-testing machine or frame, and load is applied to the upper end of the steel section. Slip between the steel member and the

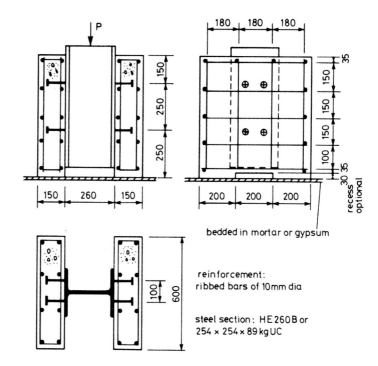

Figure 2.9   Standard push test

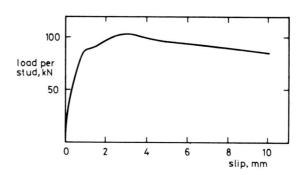

Figure 2.10   Typical load–slip curve for 19-mm stud connectors in a composite slab

two slabs is measured at several points, and the average slip is plotted against the load per connector. A typical load–slip curve is shown in Fig. 2.10, from a test using composite slabs [24].

In practice, designers normally specify shear connectors for which strengths have already been established, for it is an expensive matter to carry out sufficient tests to determine design strengths for a new type of connector. If reliable results are to be obtained, the test must be specified in detail, as the load–slip relationship is influenced by many variables, including:

(1) number of connectors in the test specimen,
(2) mean longitudinal stress in the concrete slab surrounding the connectors,
(3) size, arrangement and strength of slab reinforcement in the vicinity of the connectors,
(4) thickness of concrete surrounding the connectors,
(5) freedom of the base of each slab to move laterally, and so to impose uplift forces on the connectors,
(6) bond at the steel–concrete interface,
(7) strength of the concrete slab, and
(8) degree of compaction of the concrete surrounding the base of each connector.

The details shown in Fig. 2.9 include requirements relevant to items 1 to 6. The amount of reinforcement specified and the size of the slabs are greater than for the British standard test, which has barely changed since it was introduced in 1965. The Eurocode test gives results that are less influenced by splitting of the slabs, and so give better predictions of the behaviour of connectors in beams [17].

Tests have to be done for a range of concrete strengths, because the strength of the concrete influences the mode of failure, as well as the failure load. Studs may reach their maximum load when the concrete surrounding them fails, but in stronger concrete, they shear off. This is why the design shear resistance of studs with $h/d \geq 4$ is given in Eurocode 4 as the lesser of two values:

$$P_{Rd} = \frac{0.8 f_u (\pi d^2/4)}{\gamma_V} \tag{2.14}$$

and

$$P_{Rd} = \frac{0.29 d^2 (f_{ck} E_{cm})^{1/2}}{\gamma_V} \tag{2.15}$$

where f_u is the ultimate tensile strength of the steel (≤ 500 N/mm^2), and f_{ck} and E_{cm} are the cylinder strength and mean secant (elastic) modulus of the concrete, respectively. Dimensions h and d are shown in Fig. 2.6. The value recommended for the partial safety factor γ_V is 1.25, based on statistical calibration studies. When $f_u = 450$ N/mm^2, Equation 2.14 governs when f_{ck} exceeds about 30 N/mm^2.

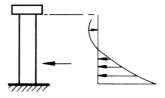

Figure 2.11   Bearing stress on the shank of a stud connector

Ignoring γ_V, it is evident that Equation 2.14 represents shear failure in the shank of the stud at a mean stress of $0.8f_u$. To explain Equation 2.15, it is assumed that the force P_R is distributed over a length of connector equal to twice the shank diameter, because research has shown that the bearing stress on a shank is concentrated near the base, as sketched in Fig. 2.11. An approximate mean stress is then $0.145\,(f_{ck}E_{cm})^{1/2}$. Its value, using data from EN 1992, ranges from 110 N/mm² for class C20/25 concrete to 171 N/mm² for class C40/50 concrete, so for these concretes the mean bearing stress at concrete failure ranges from $5.5f_{ck}$ to $4.3f_{ck}$. This estimate ignores the enlarged diameter at the weld collar at the base of the stud, shown in Fig. 2.6; but it is clear that the effective compressive strength is several times the cylinder strength of the concrete.

This very high strength is possible only because the concrete bearing on the connector is restrained laterally by the surrounding concrete, its reinforcement, and the steel flange. The results of push tests are likely to be influenced by the degree of compaction of the concrete, and even by the arrangement of particles of aggregate, in this small but critical region. This is thought to be the main reason for the scatter of the results obtained.

The usual way of allowing for this scatter is to specify that the characteristic resistance P_{Rk} be taken as 10% below the lowest of the results from three tests, and then corrected for any excess of the measured strength of the connector material above the minimum specified value.

The other property that can be deduced from a push-test result is the *slip capacity*, δ_u. This is defined in EN 1994-1-1 as the maximum slip at the load level P_{Rk}, normally on the falling branch of the load–slip curve. The characteristic slip capacity, δ_{uk}, is the minimum value of δ_u from a set of tests, reduced by 10%, unless there are sufficient test results for the 5% lower fractile to be determined.

In EN 1994-1-1, a connector may be taken as 'ductile' if $\delta_{uk} \geq 6$ mm. The use of ductile connectors leads to simpler design, as explained in Section 3.6.2.

The load–slip curve for a connector in a beam is influenced by the difference between the longitudinal stress in a concrete flange and that in the slabs in a push test. Where the flange is in compression the load/slip

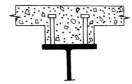

Figure 2.12   Haunch with connectors too close to a free surface

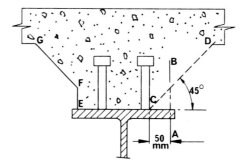

Figure 2.13   Detailing rules for haunches

ratio (the stiffness) in the elastic range exceeds the push-test value, but the ultimate strength is about the same. For slabs in tension (e.g., in a region of hogging moment), the connection is significantly less stiff [25] but the ultimate shear resistance is only slightly lower. The reduction in stiffness is one reason why partial shear connection (Section 3.6) is allowed in Eurocode 4 only in regions of sagging bending moment.

There are two situations in which the resistance of a connector found from push tests may be too high for use in design. One is repeated loading, such as that due to the passage of traffic over a bridge. The other is where the lateral restraint to the concrete in contact with the connector is less than that provided in a push test, as in a haunched beam with connectors too close to a free surface (Fig. 2.12). For this reason, the use of the standard equations for resistance of connectors is allowed in haunched beams only where the cross-section of the haunch satisfies certain conditions. In EN 1994-1-1 these are that the concrete cover to the side of the connectors may not be less than 50 mm (line AB in Fig. 2.13), and that the free concrete surface may not lie within the line CD, which runs from the base of the connector at an angle of 45° with the steel flange. A haunch that just satisfies these rules is shown as EFG.

There are also rules for the detailing of reinforcement for haunches, which apply also at the free edge of an L-beam.

Tests show that the ability of lightweight-aggregate concrete to resist the high local stresses at shear connectors is slightly less than that of normal-density concrete of the same cube strength. This is allowed for in EN 1994-1-1 by the lower value of E_{cm} that is specified for lightweight concrete (Table 1.4).

2.5.1 *Stud connectors used with profiled steel sheeting*

Where profiled sheeting is used, stud connectors are located within concrete ribs that have the shape of a haunch, which may run in any direction relative to the direction of span of the composite beam. Tests show that the shear resistance of connectors is sometimes lower than it is in a solid slab, for materials of the same strength, because of local failure of the concrete rib.

For this reason, EN 1994-1-1 specifies reduction factors, applied to the resistance P_{Rd} found from Equation 2.14 or 2.15. For sheeting with ribs parallel to the beam, the factor is

$$k_\ell = 0.6 \frac{b_0}{h_p} \left(\frac{h}{h_p} - 1 \right) \leq 1.0 \tag{2.16}$$

where the dimensions b_0, h_p and h are illustrated in Fig. 2.14, and h is taken as not greater than $h_p + 75$ mm.

For sheeting with ribs transverse to the beam the factor is

$$k_t = \frac{0.7}{\sqrt{n_r}} \frac{b_0}{h_p} \left(\frac{h}{h_p} - 1 \right) \tag{2.17}$$

where n_r is the number of connectors in one rib where it crosses a beam, not to be taken as greater than 2 in calculations. EN 1994-1-1 gives upper limits to k_t that range from 0.6 to 1.0. The limit depends on the thickness of the sheeting, the diameter of the studs, and on whether n_r is 1 or 2. For $n_r > 2$, comment is given in Section 3.11.2. The limits also distinguish between studs welded to the steel flange through a hole in the sheeting (the usual practice in some countries) and the British (and North American) practice of 'through-deck welding'.

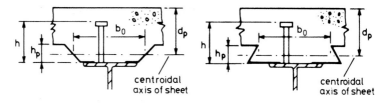

Figure 2.14 Composite beam and composite slab spanning in the same direction

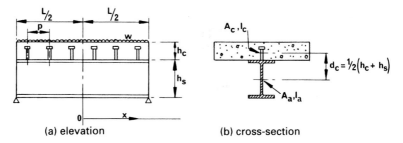

(a) elevation   (b) cross-section

Figure 2.15   Simply-supported composite beam

2.6   Partial interaction

In studying the simple composite beam with full interaction (Section 2.2.2), it was assumed that slip was everywhere zero. However, the results of push tests show (e.g., Fig. 2.10) that, even at the smallest loads, slip is not zero. It is therefore necessary to know how the behaviour of a beam is modified by the presence of slip. This is best illustrated by an analysis based on elastic theory. It leads to a differential equation that has to be solved afresh for each type of loading, and is therefore too complex for use in design offices. Even so, partial-interaction theory is useful, for it provides a starting point for the development of simpler methods for predicting the behaviour of beams at working load, and finds application in the calculation of interface shear forces due to shrinkage and differential thermal expansion.

The problem to be studied and the relevant variables are defined below. The details of the theory, and of its application to a composite beam, are given in Appendix A. The results and comments on them are given below and in Section 2.7.

Elastic analysis is relevant to situations in which the loads on connectors do not exceed about half their ultimate strength. The relevant part OB of the load–slip curve (Fig. 2.10) can be replaced with little error by the straight line OB. The ratio of load to slip given by this line is known as the *connector modulus*, k.

For simplicity, the scope of the analysis is restricted to a simply supported composite beam of span L (Fig. 2.15), carrying a distributed load w per unit length. The cross-section consists of a concrete slab of thickness h_c, cross-sectional area A_c, and second moment of area I_c, and a symmetrical steel section with corresponding properties h_s, A_a and I_a. The distance between the centroids of the concrete and steel cross-sections, d_c, is given by

$$d_c = \frac{h_c + h_s}{2} \tag{2.18}$$

Shear connectors of modulus k are provided at uniform spacing p along the length of the beam.

The elastic modulus of the steel is E_a, and that of the concrete for short-term loading is E_c. Allowance is made for creep of concrete by using an effective modulus E'_c in the analysis, where

$$E'_c = k_c E_c$$

and k_c is a reduction coefficient, calculated from the ratio of creep strain to elastic strain. The modular ratio n is defined by $n = E_a/E_c$, so that

$$E'_c = \frac{k_c E_a}{n} \qquad (2.19)$$

The concrete is assumed to be as stiff in tension as it is in compression, for it is found that tensile stresses in concrete are low enough for little error to result in this analysis, except when the degree of shear connection is very low.

The results of the analysis are expressed in terms of two functions of the cross-section of the member and the stiffness of its shear connection, α and β. These are defined by the following equations, in which notation that was established in CP117:Part 2 [26] has been used.

$$\frac{1}{A_0} = \frac{n}{k_c A_c} + \frac{1}{A_a} \qquad (2.20)$$

$$\frac{1}{A'} = d_c^2 + \frac{I_0}{A_0} \qquad (2.21)$$

$$I_0 = \frac{k_c I_c}{n} + I_a \qquad (2.22)$$

$$\alpha^2 = \frac{k}{pE_a I_0 A'} \qquad (2.23)$$

$$\beta = \frac{A'pd_c}{k} \qquad (2.24)$$

In a composite beam, the steel section is thinner than the concrete section, and the steel has a much higher coefficient of thermal conductivity. Thus the steel responds more rapidly than the concrete to changes of temperature. If the two components were free, their lengths would change at different rates; but the shear connection prevents this, and the resulting stresses in both materials can be large enough to influence design. The shrinkage of the concrete slab has a similar effect. A simple way of allowing for such differential strains in this analysis is to assume that, after connection to the steel, the concrete slab shortens uniformly, by an amount ε_c per unit length, relative to the steel.

It is shown in Appendix A that the governing equation relating slip s to distance along the beam from mid-span, x, is

$$\frac{d^2 s}{dx^2} - \alpha^2 s = -\alpha^2 \beta w x \tag{2.25}$$

and that the boundary conditions for the present problem are:

$$\left.\begin{array}{ll} s = 0 & \text{when } x = 0 \\[2ex] \dfrac{ds}{dx} = -\varepsilon_c & \text{when } x = \pm L/2 \end{array}\right\} \tag{2.26}$$

The solution of Equation 2.25 is then

$$s = \beta w x - \left(\frac{\beta w + \varepsilon_c}{\alpha}\right) \text{sech}\left(\frac{\alpha L}{2}\right) \sinh \alpha x \tag{2.27}$$

Expressions for the slip strain and the stresses throughout the beam can be obtained from this result. The stresses at a cross-section are found to depend on the loading, boundary conditions and shear connection for the whole beam. They cannot be calculated from the bending moment and shear force at the section considered. This is the main reason why design methods simple enough for use in practice have to be based on full-interaction theory.

2.7   Effect of slip on stresses and deflections

Full-interaction and no-interaction elastic analyses are given in Section 2.2 for a composite beam made from two elements of equal size and stiffness. Its cross-section (Fig. 2.2(b)) can be considered as the transformed section for the steel and concrete beam shown in Fig. 2.16. Partial-interaction analysis of this beam (Appendix A) illustrates well the effect of connector flexibility on interface slip and hence on stresses and deflections, even though the cross-section is not one that would be used in practice.

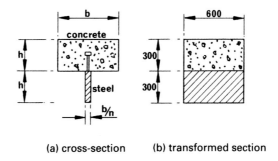

(a) cross-section   (b) transformed section

Figure 2.16   Transformed section of steel and concrete beam

The numerical values, chosen to be typical of a composite beam, are given in Section A.2. Substitution in Equation 2.27 gives the relation between s and x for a beam of depth 0.6 m and span 10 m as

$$10^4 s = 1.05x - 0.0017 \sinh(1.36x) \tag{2.28}$$

The maximum slip occurs at the ends of the span, where $x = \pm 5$ m. From Equation 2.28, it is ± 0.45 mm.

The results obtained in Sections 2.2.1 and 2.2.2 are also applicable to this beam. From Equation 2.6, the maximum slip if there were no shear connection would be ± 8.1 mm. Thus the shear connectors reduce end slip substantially, but do not eliminate it. The variations of slip strain and slip along the span for no interaction and partial interaction are shown in Fig. 2.3.

The connector modulus k was taken as 150 kN/mm (Appendix A). The maximum load per connector is k times the maximum slip, so the partial-interaction theory gives this load as 67 kN, which is sufficiently far below the ultimate strength of 100 kN per connector for the assumption of a linear load–slip relationship to be reasonable. Longitudinal strains at mid-span given by full-interaction and partial-interaction theories are shown in Fig. 2.17. The increase in extreme-fibre strain due to slip, 28×10^{-6},

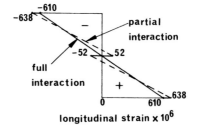

Figure 2.17   Longitudinal strains at mid-span

is much less than the slip strain at the interface, 104×10^{-6}. The maximum compressive stress in the concrete is increased by slip from 12.2 to 12.8 N/mm², a change of 5%. This higher stress is 43% of the cube strength, so the assumption of elastic behaviour is reasonable.

The ratio of the partial-interaction curvature to the full-interaction curvature is 690/610, or 1.13. Integration of curvatures along the beam shows that the increase in deflection, due to slip, is also about 13%. The effects of slip on deflection are found in practice to be less than is implied by this example, because here a rather low value of connector modulus has been used and the effect of bond has been neglected.

The longitudinal compressive force in the concrete at mid-span is proportional to the mean compressive strain. From Fig. 2.17, this is 305×10^{-6} for full interaction and 293×10^{-6} for partial interaction, a reduction of 4%.

The influence of slip on the flexural behaviour of the member may be summarised as follows. The bending moment at mid-span, $wL^2/8$, can be considered to be the sum of a 'concrete' moment M_c, a 'steel' moment M_a, and a 'composite' moment Fd_c (Fig. A.1):

$$M_c + M_a + Fd_c = \frac{wL^2}{8}$$

In the full-interaction analysis, Fd_c contributes 75% of the total moment, and M_c and M_a 12.5% each. The partial-interaction analysis shows that slip reduces the contribution from Fd_c to 72% of the total, so that the contributions from M_c and M_a rise to 14%, corresponding to an increase in curvature of $(14 - 12.5)/12.5$, or 12%.

The interface shear force per unit length, $v_{L,x}$ is given by Equation 2.12 for full interaction and by Equations A.1 and 2.28 for partial interaction. The expressions for $v_{L,x}$ over a half span are plotted in Fig. 2.18. They show that, in the elastic range, the distribution of loading on the connectors is similar to that given by full-interaction theory. The reasons for using uniform rather than 'triangular' spacing of connectors are discussed in Section 3.6.

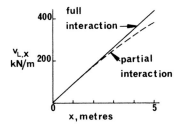

Figure 2.18 Longitudinal shear force per unit length

2.8   Longitudinal shear in composite slabs

There are three types of shear connection between a profiled steel sheet and a concrete slab. At first, reliance was placed on the natural bond between the two. This is unreliable unless separation at the interface ('uplift') is prevented, so sheets with re-entrant profiles, such as Holorib, were developed. This type of shear connection is known as 'frictional interlock'. The second type is 'mechanical interlock', provided by pressing dimples or ribs (Fig. 2.8) into the sheet. The effectiveness of these embossments depends entirely on their depth, which must be accurately controlled during manufacture. The third type of shear connection is 'end anchorage'. This can be provided where the end of a sheet rests on a steel beam, by means of shot-fired pins, or by welding studs through the sheeting to the steel flange.

2.8.1   *The* m–k *or shear-bond test*

The effectiveness of shear connection is studied by means of loading tests on simply-supported composite slabs, as sketched in Fig. 2.19. Specifications for such tests are given in clause B.3 of EN 1994-1-1. The length of each shear span, L_s, is usually $L/4$, where L is the span, which is typically about 3.0 m. There are three possible modes of failure:

- in flexure, at a cross-section such as 1–1 in Fig. 2.19,
- in longitudinal shear, along a length such as 2–2, and
- in vertical shear, at a cross-section such as 3–3.

The expected mode of failure in a test depends on the ratio of L_s to the effective depth d_p of the slab, shown in Fig. 2.20. In tests to EN 1994-1-1, the results are plotted on a diagram with axes V/bd_p and A_p/bL_s (Fig. 2.21), for reasons that are now explained.

At high L_s/d_p, flexural failure occurs. The maximum bending moment, M_u, is given by

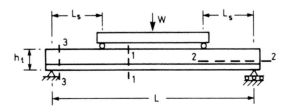

Figure 2.19   Critical sections for a composite slab

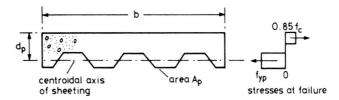

Figure 2.20 Bending resistance of a composite slab

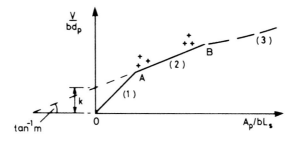

Figure 2.21 Definition of m and k

$$M_u = VL_s \tag{2.29}$$

where V is the maximum vertical shear, assumed to be much greater than the self-weight of the slab. A test specimen, of breadth b, should include a number of complete wavelengths of sheeting, of total cross-sectional area A_p. Flexural failure is modelled by simple plastic theory, with all the steel at its yield stress, f_{yp} (Fig. 2.20), and sufficient concrete at $0.85f_c$, where f_c is the cylinder strength, for longitudinal equilibrium. The lever arm is a little less than d_p, but approximately,

$$M_u \propto A_p f_{yp} d_p \tag{2.30}$$

From Equation 2.29

$$\frac{V}{bd_p} = \frac{M_u}{bd_pL_s} \propto \frac{A_p f_{yp}}{bL_s} \tag{2.31}$$

The strength f_{yp} is not varied during a series of tests, and has no influence on longitudinal shear failure. It is therefore omitted from the axes on Fig. 2.21, and Equation 2.31 shows that flexural failure should plot as a straight line through the origin, (1) in Fig. 2.21.

At low L_s/d_p, vertical shear failure occurs. The mean vertical shear stress on the concrete is roughly equal to V/bd_p. It is assumed in current codes that the ratio A_p/bL_s has little influence on its ultimate value, so vertical shear failures are represented by a horizontal line. However, Patrick & Bridge [27] have shown that this should be a rising curve, (3) in Fig. 2.21.

Longitudinal shear failures occur at intermediate values of L_s/d_p, and lie near the line

$$\frac{V}{bd_p} = m\left(\frac{A_p}{bL_s}\right) + k \tag{2.32}$$

as shown by AB on Fig. 2.21, where m and k are constants to be determined by testing. Design based on Equation 2.32 is one of the two methods given in EN 1994-1-1. The other is treated in Section 3.3.2. The present method is similar to one that has been widely used for several decades [19], known as the 'm–k method'. It is given in BS 5950:Part 4. In that method, m and k are usually defined by the equation

$$V = bd_p(f_c)^{1/2}\left[m\frac{A_p}{bL_s(f_c)^{1/2}} + k\right] \tag{2.33}$$

where f_c is the measured cylinder or cube strength of the concrete. This equation can give unsatisfactory results for m and k when f_c varies widely within a series of tests, so f_c has been omitted from Equation 2.32. A comparison of the two methods [17] has shown that this has little effect on m; but the two equations give different values for k, in different units. A value found by, for example, the method of BS 5950:Part 4 cannot be used in design to Eurocode 4; but a new value can sometimes be determined from the original test data [28].

A typical set of tests consists of a group of three, with L_s/d_p such that the results lie near point A on Fig. 2.21, and a second group with lower L_s/d_p, such that the results lie near point B. Values of m and k are found for a line drawn below the lowest result in each group, at a distance that allows for the scatter of the test data.

All six failures have to be in longitudinal shear. These failures typically commence when a crack occurs in the concrete under one of the load points, associated with loss of bond along the shear span and measurable slip at the end of the span. If this leads to failure of the slab, the shear connection is classified as 'brittle'. Such failures occur suddenly, and are penalised in design to EN 1994-1-1 by a 20% reduction in design resistance.

Where the eventual failure load exceeds the load causing a recorded end slip of 0.1 mm by more than 10%, the failure is classified as 'ductile'. Recently-developed profiles for sheeting have better mechanical interlock than earlier shapes, which relied more on frictional interlock and were more susceptible to 'brittle' failure. The influence of bond is minimised, in the standard test, by the application of several thousand cycles of repeated loading up to 60% of the expected failure load, before loading to failure.

When a new profile is developed, values of m and k have to be determined, in principle, for each thickness of sheeting, each overall depth of slab to be used, and for a range of concrete strengths. Codes allow some simplification, but the testing remains a long and costly process. The $m-k$ test is also unsatisfactory in other ways [17].

Chapter 3
Simply-supported composite slabs and beams

3.1 Introduction

The subjects of this and subsequent chapters are treated in the sequence in which they developed. Relevant structural behaviour is discovered by experience or research, and is then represented by mathematical models. These make use of standardised properties of materials, such as the yield strength of steel, and enable the behaviour of a member under load to be predicted. The models are developed into design rules, as found in codes of practice, by simplifying them wherever possible, defining their scope and introducing partial factors.

Research workers often propose alternative models, and language barriers are such that the model preferred in one country may be little known elsewhere. The writers of codes try to select the most rational and widely-applicable of the available models, but must also consider existing design practices and the need for simplicity. The design rules used in this book are taken from the Eurocodes, which differ slightly from the correspond-ing British codes; but the underlying models are usually the same, and significant differences will be explained.

The methods to be described are illustrated by the design calculations for part of a framed structure for a building. To avoid repetition, the results obtained at each stage are used in subsequent work.

The notation used is that explained and listed in the section 'Symbols, terminology and units'.

3.2 Example: layout, materials and loadings

In a framed structure for a wing of a building, the columns are arranged at 4 m centres in two rows 9 m apart. A design is required for a typical floor, which consists of a composite floor slab supported by, and composite with, steel beams that span between the columns as shown in Fig. 3.1.

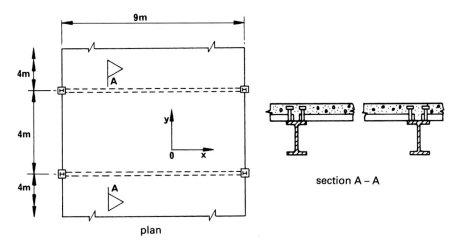

Figure 3.1   Design example – structure for a typical floor

Properties of concrete

Lightweight-aggregate concrete grade LC25/28 will be used for the floor slabs and normal-density concrete grade C25/30 for the encasement of the columns and the webs of the beams. The properties of these concretes are given in Table 1.4. The design compressive strengths for both of these concretes are:

$$f_{cd} = f_{ck}/\gamma_C = 25/1.5 = 16.7 \text{ N/mm}^2$$

The short-term elastic moduli and modular ratios are:

LC25/28: $E_{cm} = 20.7$ kN/mm^2; $n_0 = 210/20.7 = 10.1$
C25/30: $E_{cm} = 31.0$ kN/mm^2; $n_0 = 210/31 = 6.8$

The long-term compressive strain of these concretes under permanent loads is about three times the initial elastic strain, due to creep. In elastic analysis, this would require separate calculations for permanent and variable loads. For buildings, EN 1994 permits the simplification that all strains may be assumed to be twice their short-term value. This is done by using modular ratios $n = 2n_0$. Hence,

for grade LC25/28, $n = 20.2$;   for grade C25/30, $n = 13.6$

Properties of other materials

The partial factors γ_M to be used are those recommended in the Eurocodes. A national annex may prescribe other values. The following grades of the materials are those widely used:

structural steel: yield strength $f_y = f_{yd} = 355$ N/mm^2 ($\gamma_A = 1.0$)
profiled steel sheeting: yield strength $f_{yp} = f_{yp,d} = 350$ N/mm^2 ($\gamma_A = 1.0$)
reinforcement: yield strength $f_{sk} = 500$ N/mm^2, $f_{sd} = 435$ N/mm^2 ($\gamma_S = 1.15$)
welded fabric: yield strength $f_{sk} = 500$ N/mm^2, $f_{sd} = 435$ N/mm^2 ($\gamma_S = 1.15$)
shear connectors; 19-mm headed studs 100 mm high;
    ultimate strength $f_u = 500$ N/mm^2, $f_{ud} = 400$ N/mm^2 ($\gamma_V = 1.25$)

The elastic modulus for structural steel is $E_a = 210$ kN/mm^2. In design of beams and slabs, the value for reinforcement is assumed, for simplicity, also to be 210 kN/mm^2; but the more accurate value, $E_s = 200$ kN/mm^2 is used in column design.

Resistance of the shear connectors
The design shear resistance is given by Equation 2.15 and is

$$P_{Rd} = 0.29 \times 19^2 \, (25 \times 20\,700)^{0.5}/(1.25 \times 1000) = 60.2 \text{ kN} \qquad (3.1)$$

Equation 2.14 gives the higher value 91 kN, and so does not apply.

Permanent actions
From Table 1.4, the unit weights of the concretes, including reinforcement, are:

    for LC25/28, 19.5 kN/m^3;   for C25/30, 25.0 kN/m^3

For design of formwork, each value is increased by 1 kN/m^3 (from EN 1991-1-1) to allow for the higher moisture content of fresh concrete.

The unit weight of structural steel is taken as 77 kN/m^3.
The characteristic weight of floor and ceiling finishes is taken as 1.3 kN/m^2.

Variable actions
The floors to be designed are assumed to be in category C3 of EN 1991-1-1: [13] 'Areas where people may congregate, without obstacles for moving people' (e.g., exhibition rooms, etc.). The characteristic loadings are:

$$q_k = 5.0 \text{ kN/m}^2 \text{ on the whole floor area or any part of it} \qquad (3.2)$$

or

$$Q_k = 4.0 \text{ kN on any area 50 mm square} \qquad (3.3)$$

The load q_k is high for a building not intended for storage or industrial use. Its use here enables many aspects of design to be illustrated. For comparison, typical imposed loads for office areas are given by Equations 1.12 and 1.13. An allowance of $1.2 \, \text{kN/m}^2$ for non-structural partition walls is treated as an additional imposed load, because their positions are unknown.

3.3 Composite floor slabs

Composite slabs have for many decades been the most widely used method of suspended floor construction for steel-framed buildings in North America. Within the last thirty years there have been many advances in design procedures, and a wide range of profiled sheetings has become available in Europe. The British Standard for the design of composite floors [19] first appeared in 1982. There are Eurocodes for design of both the sheeting alone [15] and the composite slab [3].

The steel sheeting has to support not only the wet concrete for the floor slab, but other loads that are imposed during concreting. These may include the heaping of concrete and pipeline or pumping loads. For construction loading, EN 1991-1-6 recommends a distributed loading between 0.75 and $1.5 \, \text{kN/m}^2$. The loading used here is:

$$q_k = 1.0 \, \text{kN/m}^2$$

Profiled steel sheeting

The sheeting is very thin for economic reasons, usually between 0.8 mm and 1.2 mm. It has to be galvanised to resist corrosion, and this adds about 0.04 mm to the overall thickness. It is specified in EN 1993-1-3 that where design is based on the nominal thickness of the steel, the sheet must have at least 95% of that thickness – but it is not a simple matter for the user to check this. The sheets are pressed or cold rolled and are typically about 1-m wide and up to 6-m long. They are designed to span in the longitudinal direction only. For many years, sheets were typically 50-mm deep and the limiting span was about 3 m. The cost of propping the sheets during concreting, to reduce deflections, led to the development of deeper profiles; but design of composite slabs is still often governed by a limit on deflection.

The local buckling stress of a flat panel within sheeting should ideally exceed its yield strength; but this requires breadth/thickness ratios of less than about 35. Modern profiles have local stiffening ribs, but it is difficult to achieve slendernesses less than about 50, so that for flexure, the sections are in Class 4 (i.e., the buckling stress is below the yield stress). Calculation of the resistance to bending then becomes complex and involves iteration.

The specified or nominal yield strength is that of the flat sheet from which the sheeting is made. In the finished product, the yield strength is higher at every bend and corner, because of work hardening.

To enable it to fulfil its second role, as reinforcement for the concrete slab, dimples are pressed into the surface of the sheeting, to act as shear connectors. These dimpled areas may not be fully effective in resisting longitudinal stress, so both they and the local buckling reduce the second moment of area (I) of the sheeting to below the value calculated for the gross steel section.

For these reasons, manufacturers commission tests on prototype sheets, and provide designers either with test-based values of resistance and stiffness, or with 'safe load' tables calculated from those values.

Design of composite slab

The cross-sectional area of steel sheeting that is needed for the construction phase often provides more than enough bottom reinforcement for the composite slab. It is then usual to design the slabs as simply-supported. The concrete is of course continuous over the supporting beams, and the sheets may be as well (e.g., if 6-m sheets are used for a succession of 3-m spans).

These 'simply-supported' slabs require top longitudinal reinforcement at their supports, to control the widths of cracks. The amount is specified in EN 1994-1-1 as 0.2% of the cross-sectional area of concrete above the steel ribs for unpropped construction and 0.4% for propped construction.

Long-span slabs are sometimes designed as continuous over their supports. They are analysed as described in Section 4.7. Several action effects that have to be considered in the design of composite slabs are now considered. The methods are illustrated by the worked example in Section 3.4.

3.3.1 *Resistance of composite slabs to sagging bending*

The width of slab considered in calculations, b, is usually taken as one metre, but for clarity only a width of one wavelength is shown in Fig. 3.2. The overall thickness h is required by EN 1994-1-1 to be not less than 80 mm; and the thickness of concrete above the 'main flat surface' of the top of the ribs of the sheeting, to be not less than 40 mm. Normally, this thickness is 60 mm or more, to provide sufficient sound or fire insulation, and resistance to concentrated loads.

Except where the sheeting is unusually deep, the neutral axis for bending lies in the concrete, where there is full shear connection; but in regions with partial shear connection, there is usually a second neutral axis within the steel section. Local buckling of compressed sheeting then has to be considered. This is done by using effective widths for flat regions of

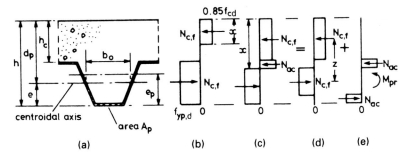

Figure 3.2 Cross-section of composite slab, and stress blocks for sagging bending

sheeting. These widths are allowed (in EN 1994-1-1) to be up to twice the limits given for Class 1 steel web plates in beams, because the concrete prevents the sheeting from buckling upwards, which shortens the wavelength of the buckles.

For sheeting in tension, the width of embossments should be neglected in calculating the effective area, unless tests have shown that a larger area is effective.

For these reasons, the effective area of width b of sheeting, A_p, and the height of the centre of area above the bottom of the sheet, e, are usually based on tests. These usually show that e_p, the height of the plastic neutral axis of the sheeting, is different from e.

Because local buckling is allowed for in this way, the bending resistance of width b of composite slab can be calculated by simple plastic theory. There are three situations, as follows.

(1) Neutral axis above the sheeting

The assumed distribution of longitudinal bending stresses is shown in Fig. 3.2(b). There must be full shear connection, so that the design compressive force in the concrete, $N_{c,f}$, is equal to the yield force for the steel:

$$N_{c,f} = A_p f_{yp,d} \tag{3.4}$$

where $f_{yp,d}$ is the design yield strength of the sheeting. The depth of the stress block in the concrete is given by

$$x = x_{pl} = N_{c,f}/(0.85 f_{cd} b) \tag{3.5}$$

For simplicity, and consistency with the method for composite beams, the depth to the neutral axis is assumed also to be x_{pl}, even though this is not in accordance with EN 1992. This method is therefore valid when

$$x_{pl} \leq h_c$$

and gives

$$M_{Rd} = N_{c,f}(d_p - 0.5x_{pl}) \tag{3.6}$$

where M_{Rd} is the design resistance to sagging bending.

(2) Neutral axis within the sheeting, and full shear connection

The stress distribution is shown in Fig. 3.2(c). The force $N_{c,f}$ is now less than that given by Equation 3.4 and is

$$N_{c,f} = 0.85f_{cd}bh_c \tag{3.7}$$

because compression within ribs is neglected, for simplicity. There is a compressive force N_{ac} in the sheeting. There is no simple method of calculating x or the force N_{ac}, because of the complex properties of profiled sheeting, so the following approximate method is used. The tensile force in the sheeting is decomposed, as shown in Figs 3.2(d) and (e), into a force at the bottom equal to N_{ac} (the compressive force) and a force N_p, where

$$N_p = N_{c,f} \tag{3.8}$$

The equal and opposite forces N_{ac} provide a resistance moment M_{pr}, equal to the resistance moment for the sheeting, M_{pa}, reduced by the effect of the axial force $N_{c,f}$. It should be noted that, in EN 1994-1-1, the value represented by the symbol $N_{c,f}$ depends on the ratio x/h_c. It is the lesser of the two values given by Equations 3.4 and 3.7. This can be confusing; so for clarity here, a further symbol, N_{pa}, is introduced. It always has the value

$$N_{pa} = A_p f_{yp,d} \tag{3.9}$$

The subscript f in $N_{c,f}$ denotes full shear connection. Where there is partial shear connection, the compressive force in the concrete slab is N_c, which cannot exceed $N_{c,f}$.

The relationship between M_{pr}/M_{pa} and $N_{c,f}/N_{pa}$ depends on the profile, but is typically as shown by the dashed curve ABC in Fig. 3.3(a). This is approximated in Eurocode 4 by the equation

$$M_{pr} = 1.25M_{pa}[1 - (N_{c,f}/N_{pa})] \leq M_{pa} \tag{3.10}$$

which is shown as ADC. The resistance moment is then given by

$$M_{Rd} = N_{c,f}z + M_{pr} \tag{3.11}$$

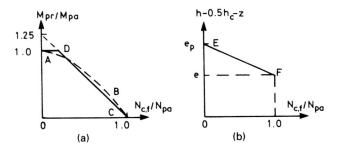

Figure 3.3   Equations 3.10 and 3.14

as shown in Fig. 3.2(d) and (e). The lever arm z is found by the approximation shown by line EF in Fig. 3.3(b). This is clearly correct when $N_{c,f} = N_{pa}$, because N_{ac} is then zero, so M_{pr} is zero. Equation 3.6 with $x_{pl} = h_c$ then gives M_{Rd}. The lever arm is

$$z = d_p - 0.5h_c = h - e - 0.5h_c \qquad (3.12)$$

as given by point F.

To check point E, we assume that $N_{c,f}$ is nearly zero (e.g., if the concrete is very weak), so that $M_{pr} \simeq M_{pa}$. The neutral axis for M_{pa} alone is at height e_p above the bottom of the sheet, and the lever arm for the force $N_{c,f}$ is

$$z = h - e_p - 0.5h_c \qquad (3.13)$$

as given by point E. This method has been validated by tests.

The line EF is given by

$$z = h - 0.5h_c - e_p + (e_p - e)N_{c,f}/N_{pa} \qquad (3.14)$$

(3) Partial shear connection

The compressive force in the slab, N_c, is now less than $N_{c,f}$ and is determined by the strength of the shear connection. The depth x of the stress block is given by

$$x = N_c/(0.85f_{cd}b) \le h_c \qquad (3.15)$$

There is a second neutral axis within the steel sheeting. The stress blocks are as shown in Fig. 3.2(b) for the slab (with force N_c, not $N_{c,f}$), and Fig. 3.2(c) for the sheeting. The calculation of M_{Rd} is as for method (2), except that $N_{c,f}$ is replaced by N_c, and h_c by x, so that:

$$z = h - 0.5x - e_p + (e_p - e)N_c/N_{pa} \qquad (3.16)$$

$$M_{pr} = 1.25M_{pa}[1 - (N_c/N_{pa})] \leq M_{pa} \tag{3.17}$$

$$M_{Rd} = N_c z + M_{pr} \tag{3.18}$$

3.3.2 *Resistance of composite slabs to longitudinal shear*

For profiled sheeting that relies on frictional interlock to transmit longitudinal shear, there is no satisfactory conceptual model. This led to the development of the shear-bond test, described in Section 2.8.1, and the empirical '*m–k*' method of design, where the shear resistance is given by an equation based on Equation 2.33, in the British code [19], or on Equation 2.32, in EN 1994-1-1. With the safety factor added, the Eurocode equation is

$$V_{\ell,Rd} = bd_p[mA_p/(bL_s) + k]/\gamma_{Vs} \tag{3.19}$$

where m and k are constants with dimensions of stress, determined from shear-bond tests, and $V_{\ell,Rd}$ is the design *vertical* shear resistance for a width of slab b. This must exceed the vertical shear at an end support at which longitudinal shear failure could occur in a shear span of length L_s, shown by line 2–2 in Fig. 2.19.

For uniformly-distributed load on a span L, the length L_s is taken as $L/4$. The principle that is used when calculating L_s for other loadings is now illustrated by an example.

Calculation of L$_s$
The composite slab shown in Fig. 3.4(a) has a distributed load w per unit length and a centre point load wL, so the shear force diagram is as shown

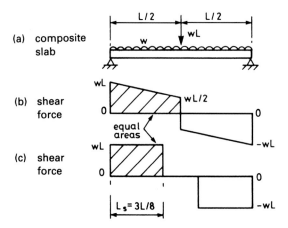

Figure 3.4 Calculation of L_s for composite slab

in Fig. 3.4(b). A new shear force diagram is constructed for a span with two point loads only, and the same two end reactions, such that the areas of the positive and negative parts of the diagram equal those of the original diagram. This is shown in Fig. 3.4, in which each shaded area is $3wL^2/8$. The positions of the point loads define the lengths of the shear spans. Here, each one is $3L/8$.

Defects of the m–k method

The method has proved to be an adequate design tool for profiles with short spans and rather brittle behaviour, which have been widely used in North America. However, to exploit fully the ductile behaviour of profiles now available, with good mechanical interlock and longer spans, it is necessary to use a partial-interaction method, as explained below.

The defects of the $m-k$ method and of profiles with brittle behaviour are given in papers that set out the new methods, by Bode & Sauerborn in Germany [29] and by Patrick & Bridge in Australia [30]. They are as follows.

(1)   The $m-k$ method is not based on a mechanical model, so that conservative assumptions have to be made in design when the dimensions, materials or loading differ from those used in the tests. The calculation of L_s, above, is an example of this.

(2)   Many additional tests are needed before the range of application can be extended; for example, to include end anchorage or the use of longitudinal reinforcing bars.

(3)   The method of evaluation of test data is the same, whether the failure is brittle or ductile. The use in EN 1994-1-1 of a penalty factor of 0.8 for brittle behaviour does not adequately represent the advantage of using sheeting with good mechanical interlock, because the advantage increases with span.

(4)   The method does not allow correctly for the beneficial effect of friction above supports, which is greater in short shear spans.

Partial-interaction design

This method is based on results from shear-bond tests [29]. For composite slabs of given cross-section and materials, the result of each test on a profile with ductile behaviour enables the degree of partial shear connection in that test to be calculated. This gives the compressive force, N_c, transferred from the sheeting to the slab within the shear span of known length, L_s. It is assumed that, before maximum load is reached, there is complete redistribution of longitudinal shear stress at the interface, so a value for the mean ultimate shear stress τ_u can be calculated. This is done for a range of shear spans, and the lowest τ_u thus found is the basis for a design value, $\tau_{u,Rd}$. (This is where the greater effect of friction in short spans is neglected.)

At an end support, the bending resistance of the slab is that of the sheeting alone (unless it is enhanced by the use of end anchorage, as described later). At any cross-section at distance x from the support, the compressive force in the slab can be calculated from $\tau_{u,Rd}$. This force may optionally be increased by μR_{Ed}, where μ is a coefficient of friction and R_{Ed} is the support reaction. The partial-interaction method of Section 3.3.1(3) enables the bending resistance, M_{Rd}, at that cross-section to be calculated. There may be a mid-span region where full shear connection is achieved and M_{Rd} is independent of x.

For safe design, this curve of M_{Rd} as a function of x (the resistance diagram) must at all points lie above the bending-moment diagram for the applied loading. If the loading is increased until the curves touch, the position of the point of contact gives the location of the cross-section of flexural failure and, if the interaction is partial, the length of the shear span.

The resistance diagram can easily be modified to take advantage of any end anchorage or slab reinforcement, and the loading diagram can be of any shape.

A worked example using data from shear-bond tests is given in Section 3.4.3.

The only type of end anchorage for which design rules are given in British or European codes is the headed stud, welded through the sheeting to the top flange of a steel beam. The resistance of the anchorage is based on local failure of the sheeting, as explained elsewhere [17].

3.3.3   *Resistance of composite slabs to vertical shear*

Tests show that resistance to vertical shear is provided mainly by the concrete ribs. For open profiles, their effective width b_0 should be taken as the mean width, though the width at the centroidal axis (Fig. 3.2(a)) is accurate enough. For re-entrant profiles, the minimum width should be used.

This shear resistance is given by the method of EN 1992-1-1 for concrete beams. Reinforcement contributes to the resistance only where it is fully anchored beyond the cross-section considered. The sheeting is unlikely to satisfy this condition. The resistance of a composite slab with ribs of effective width b_0 at spacing b is then

$$V_{Rd} = (b_0/b)d_p v_{min} \text{ per unit width} \tag{3.20}$$

where d_p is the depth to the centroidal axis (Fig. 3.2(a)) and v_{min} is the shear strength of the concrete.

The recommended value for v_{min} is

$$v_{min} = 0.035[1 + (200/d_p)^{1/2}]^{3/2} f_{ck}^{1/2} \tag{3.21}$$

with d_p in mm, taken as not less than 200, and v_{min} and f_{ck} in N/mm² units. The expression $[1 + (200/d_p)^{1/2}]^{3/2}$ allows approximately for the reduction in shear strength of concrete that occurs as the effective depth increases.

In reality, the shear stress in the side walls of the steel troughs may be quite high during the construction phase. This can be ignored when checking the composite slab, and V_{Ed} should be taken as the whole of the vertical shear, including that initially resisted by the sheeting.

Resistance to vertical shear is most likely to be critical in design where span/depth ratios are low, as is the case for beams.

3.3.4 *Punching shear*

Where a thin composite slab has to be designed to resist point loads (e.g., from a steel wheel of a loaded trolley), resistance to punching shear should be checked. Failure is assumed to occur on a 'critical perimeter', of length c_p, which is defined as for reinforced concrete slabs. For a loaded area a_p by b_p, remote from a free edge, and 45° spread through a screed of thickness h_f, it is as shown in Fig. 3.5(a):

$$c_p = 2\pi h_c + 2(b_p + 2h_f) + 2(a_p + 2h_f + 2d_p - 2h_c) \tag{3.22}$$

A reinforcing mesh is likely to be present above the sheeting. Let its areas of steel be $A_{s,x}$ and $A_{s,y}$, per unit width of slab. The effective depth of the slab may be taken as h_c (Fig. 3.5(b)), giving the reinforcement ratios as $\rho_x = A_{s,x}/h_c$ and $\rho_y = A_{s,y}/h_c$. The effective ratio is given in EN 1992-1-1 as

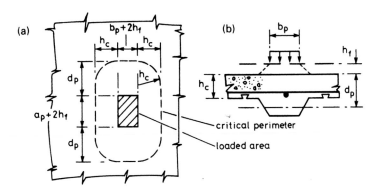

Figure 3.5   Critical perimeter for punching shear

$$\rho = (\rho_x \rho_y)^{1/2} \le 0.02$$

and the design shear stress as

$$v_{Rd} = (0.18/\gamma_C)[1 + (200/d)^{1/2}](100\rho f_{ck})^{1/3} \ge v_{min} \tag{3.23}$$

where:

v_{min} is given by Equation 3.21,
d is the mean of the effective depths of the two layers of reinforcement, but not less than 200 mm,
γ_C has the recommended value 1.50, and the units are as for Equation 3.21.

The punching shear resistance is

$$V_{Rd} = v_{Rd}c_p d \tag{3.24}$$

It is not clear from EN 1994-1-1 whether account can be taken of contributions from the concrete ribs and the sheeting. None has been assumed here, so Equation 3.24 is likely to give a conservative result.

3.3.5 *Bending moments from concentrated point and line loads*

Since composite slabs span in one direction only, their ability to carry masonry partition walls or other heavy local loads is limited. Rules are given in EN 1994-1-1 (and in the British code) for widths of composite slabs effective for bending and vertical shear resistance, for point and line loads, as functions of the shape and size of the loaded area. These are based on a mixture of simplified analyses, test data and experience.

Where transverse reinforcement is provided with a cross-sectional area of at least 0.2% of the area of concrete above the ribs of the sheeting, no calculations are needed for characteristic concentrated loads not exceeding 7.5 kN.

The rules for use where this simplification does not apply are now explained, with reference to a rectangular loaded area a_p by b_p, with its centre distance L_p from the nearer support of a slab of span L, as shown in Fig. 3.6(a). The load may be assumed to be distributed over a width b_m, defined by lines at 45° (Fig. 3.6(b)), where

$$b_m = b_p + 2(h_f + h_c) \tag{3.25a}$$

and h_f is the thickness of finishes, if any. The code does not refer to distribution in the spanwise direction, but it would be reasonable to use the same rule, and take the loaded length as

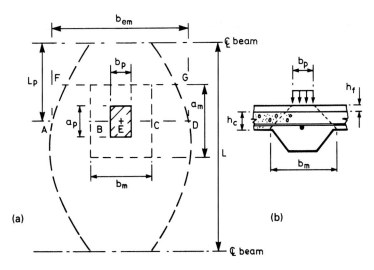

Figure 3.6 Effective width of composite slab, for concentrated load

$$a_m = a_p + 2(h_f + h_c) \tag{3.25b}$$

The width of slab assumed to be effective for global analysis and for resistance is given by

$$b_{em} = b_m + kL_p[1 - (L_p/L)] \le \text{width of slab} \tag{3.26}$$

where k is taken as 2 for bending and longitudinal shear (except for interior spans of continuous slabs, where $k = 1.33$) and as 1 for vertical shear. These rules become unreliable where the depth of the ribs is a high proportion of the total thickness. Their use is limited in EN 1994-1-1 to slabs with $h_p/h \le 0.6$.

For a simply-supported slab of span L and a point load Q_{Ed}, the sagging moment per unit width of slab on line AD in Fig. 3.6(a) is thus

$$m_{Ed} = Q_{Ed}L_p[1 - (L_p/L)]/b_{em} \tag{3.27}$$

which is a maximum when $L_p = L/2$.

The variation of b_{em} with L_p is shown in Fig. 3.6(a). The load is assumed to be uniformly-distributed along line BC, whereas the resistance is distributed along line AD, so there is sagging transverse bending under the load. The maximum sagging bending moment is at E, and is given by

$$M_{Ed} = Q_{Ed}(b_{em} - b_m)/8 \tag{3.28}$$

The sheeting has no tensile strength in this direction because the corrugations can open out, so bottom reinforcement (Fig. 3.6(b)) must be provided.

It is suggested that this reinforcement should be spread over the length a_m given by Equation 3.25b.

Vertical shear should be checked along a line such as FG, when L_p is such that FG is above the edge of the flange of the steel beam. It rarely governs design.

3.3.6 *Serviceability limit states for composite slabs*

Cracking of concrete

The lower surface of the slab is protected by the sheeting. Cracking will occur in the top surface where the slab is continuous over a supporting beam, and will be wider if each span of the slab is designed as simply-supported, rather than continuous, and if the spans are propped during construction.

For these reasons, longitudinal reinforcement should be provided above internal supports. The minimum amounts are given by EN 1994-1-1 as 0.2% of the area of concrete above the sheeting for unpropped construction, and 0.4% if propping is used. These amounts may not ensure that crack widths do not exceed 0.3 mm. If the environment is corrosive (e.g., de-icing salt on the floor of a parking area), the slabs should be designed as continuous, with cracking controlled in accordance with EN 1992-1-1.

Deflection

Where composite slabs are designed as simply-supported and are not hidden by false ceilings, deflection may govern design. The maximum acceptable deflection is more a matter for the client than the designer; but if predicted deflections are large, the designer may have to allow for the extra weight of concrete in floors that are cast with a horizontal top surface, and provide clearance above non-structural partitions.

Limiting deflection/span ratios are given, for guidance, in national annexes. Both the ratios and the load combination for which the deflection is calculated depend on whether the deflection is reversible or not, and whether brittle finishes or partitions are at risk. The deflections of the beams that support the composite slab are also relevant.

It is known from experience that deflections are not excessive when span-to-depth ratios are kept within certain limits. In EN 1994-1-1, calculations of deflections of composite slabs may be omitted if both:

- the degree of shear connection is such that end slip does not occur under service loading, and
- the ratio of span to effective depth is below a limit given in EN 1992-1-1. The recommended limit for a simply-supported slab is 20, but this can be modified by a national annex.

The provision of anti-crack reinforcement, as specified above, should reduce deflection by a useful amount. For internal spans, the Eurocode recommends that the second moment of area of the slab should be taken as the mean of values calculated for the cracked and uncracked sections. Some of these points are illustrated in the worked example in Section 3.4.

3.3.7 *Fire resistance*

All buildings are vulnerable to damage from fire, which is usually the first accidental design situation to be considered in design, and is the only one treated in this book.

In the worked example, design is in accordance with the Eurocodes. The concepts and methods used are now introduced. Italic print is used for terms that are defined in EN 1991-1-2, *Actions on structures exposed to fire* [13] or in EN 1994-1-2, *Structural fire design* [16].

Buildings are divided by *separating members* into *fire compartments*. A fire is assumed to be confined to one compartment only, which must be designed to contain it for a specified *failure time* (or *fire resistance time*) when subjected to a given temperature–time environment or fire exposure. A *standard fire exposure* is given in EN 1991-1-2, and other curves are available that depend on the *fire load density* (calorific energy per unit area, for complete combustion of all combustible materials) within the compartment considered. These temperature–time curves are reproduced in furnaces for fire testing, and are simplified models of the effects of real fires.

The walls, floor and ceiling that enclose a compartment must have a *separating function*. This is defined using two criteria:

- *thermal insulation criterion*, denoted I, concerned with limiting the transmission of heat by conduction, and
- *integrity criterion*, denoted E, concerned with preventing the passage of flames and hot gases into an adjacent compartment.

The structure of a compartment must have a *load bearing function*, denoted R (resistance), to ensure that it can resist the design effects of actions specified for the fire situation, including the effects of thermal expansion, for a period not less than the specified failure time. The *fire resistance class* of a member or compartment is denoted (for example) R60, which means that its failure time is not less than 60 minutes.

Criterion I is met mainly by specifying minimum thicknesses of incombustible insulating materials. These also contribute to meeting criterion E, which has structural implications as well. For example, excessive thermal hogging of a beam heated from below can create a gap between it and a separating wall below.

Codes give limits to the temperature rise of non-exposed surfaces, relevant to criterion I, and detailing rules relevant to criteria I and E. Design calculations relate mainly to criterion R, and only these are further discussed in this book.

It will be seen that, for practicable design, it is necessary to simplify the predictions of both the action effects and the resistances, to a greater extent than for 'cold' design. This last term refers to the normal design for persistent situations and ultimate limit states.

3.3.7.1 *Partial safety factors for fire*

It is recommended in EN 1990 that all factors γ_F for accidental actions should be 1.0; i.e., that these actions should be so defined that $\gamma_{F,fi} = 1.0$ is the appropriate factor. Subscript fi means 'fire'.

For materials, design is based essentially on characteristic or nominal properties; so for most materials, and for shear connection, $\gamma_{M,fi} = 1.0$. The reduction below γ_M for persistent design situations is then significant for concrete ($1.5 \rightarrow 1.0$) and zero for structural steel ($1.0 \rightarrow 1.0$).

3.3.7.2 *Design action effects for fire*

For a structural member with one type only of permanent loading and no prestress, the combination for accidental design situations given in EN 1990 [12] simplifies to:

$$G_k + A_d + \psi_{1,1}Q_{k,1} + \sum_{i>1}\psi_{2,i}Q_{k,i} \tag{3.29}$$

where A_d is the design value of the accidental action, and other symbols are as in Section 1.3.2.

A floor structure for a building is usually designed for distributed loads g_k and q_k and, for fire, A_d can be taken as zero. For beams and slabs, the 'simply-supported' moments and shears are proportional to the total load per unit area. To avoid additional global analyses for fire, the action effects $E_{fi,d}$ are related to those for cold design for ultimate limit states by

$$E_{fi,d} = \eta_{fi}E_d \tag{3.30}$$

where, in the simple case considered here, η_{fi} depends on the ratio q_k/g_k and the values of γ_F and ψ_1 as follows. From Equations 3.29 and 3.30, and Expression 1.6 with $Q_{k,2} = 0$, and $\psi_1 = 0.7$ (Table 1.3):

$$\eta_{fi} = \frac{\gamma_{GA}g_k + \psi_1 q_k}{\gamma_G g_k + \gamma_Q q_k} = \frac{1.0 + 0.7(q_k/g_k)}{1.35 + 1.5(q_k/g_k)} \tag{3.31}$$

The value of η_{fi} falls from 0.74 to 0.55 as the ratio q_k/g_k rises from zero to 2.0. A note in EN 1994-1-2 gives, as a simplification, the recommended value $\eta_{fi} = 0.65$.

It often occurs in cold design that the resistance provided, R_d, exceeds the relevant action effect, E_d. This is allowed for in fire design as follows. A resistance ratio $\eta_{fi,t}$ is calculated from

$$\eta_{fi,t} = \eta_{fi}(E_d/R_d) \tag{3.32}$$

Using Equation 3.30, the verification condition $(R_{fi,d} \geq E_{fi,d})$ then becomes

$$R_{fi,d} \geq \eta_{fi}E_d = \eta_{fi,t}R_d \tag{3.33}$$

This enables tabulated design data to be presented in terms of $\eta_{fi,t}$, which typically lies between 0.3 and 0.7.

3.3.7.3 *Thermal properties of materials*

Stress–strain properties for concrete and reinforcement, and for steel are given as functions of temperature, θ, in ENs 1992-1-2 and 1993-1-2, respectively. These are used as required in the worked example.

3.3.7.4 *Design methods for resistance to fire*

The three methods given in EN 1994-1-2 are outlined below. The first (given last in the code) has the widest scope, but is the most complex. It is primarily a research tool, used to validate simpler methods.

(1) Advanced calculation models

These methods rely mainly on finite-element computations, which are done in three stages, starting with a given structure, materials and fire exposure.

(a) The theory of heat transfer is used to obtain distributions of temperature, θ, throughout the structure as functions of the time, t, since the start of the fire.

(b) From the temperatures, distributions of thermal strains and of the stiffness and strength of the materials throughout the structure are computed, for various times, t.

(c) The design resistances of the structure are computed at various times, t, using data from stage (b). These resistances diminish as t increases, and eventually one of them falls below the corresponding design action effect. The structure satisfies criterion R if the time when this occurs exceeds the specified failure time.

(2) Simple calculation models

These methods enable the three preceding stages to be applied, in simplified form, in checks on the resistances of cross-sections. These are normally done only for the temperature distribution at the specified failure time, assuming that beams and slabs are simply-supported and columns are pin-ended at each floor level. The model for a composite slab is explained in Section 3.3.7.5.

(3) Tabulated data

For cross-sections of beams and columns that are often used in practice, results of calculations by method (1) or (2) are presented in EN 1994-1-2 as tabulated values of minimum dimensions, areas of reinforcement, etc., for each fire resistance class. Methods of this type are used for the beams and columns of the worked example in this book.

3.3.7.5 *Simple calculation model for unprotected composite slab*

It is assumed that the dimensions and properties of materials for the slab are known, and that its cold design was for distributed loading on simply-supported spans, for which the bending moments R_d and E_d are known, so that $\eta_{fi,t}$ (Equation 3.32) is known.

It is assumed that the required fire resistance period ($t_{fi,d}$) is 60 minutes, and that the profiled sheeting, not protected by insulation, is heated from below by the standard fire.

Thermal insulation criterion

An equation in EN 1994-1-2 gives the fire resistance time for thermal insulation, t_i, as a function of the dimensions of the cross-section, defined in Fig. 3.7, and the density of the concrete. For the worked example that follows, it gives $t_i = 157$ minutes, so the slab can prevent fire spreading to the floor above, provided that it does not collapse. This criterion is not considered further.

Load bearing function

For the bending resistance of the slab, the strength of the steel sheeting after 60 minutes' exposure is very low, and the tensile strength of the

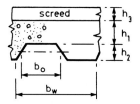

Figure 3.7 Dimensions of composite slab

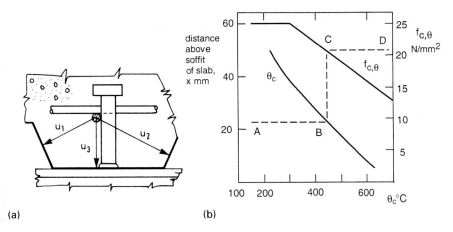

(a)   (b)

Figure 3.8 Design data for a composite slab in fire

concrete is ignored, so reinforcing bars have to be provided within the ribs. Their temperature, and hence their yield strength, depends on their effective distance from the hot surfaces, represented by z, defined by

$$z = 1/\sqrt{u_1} + 1/\sqrt{u_2} + 1/\sqrt{u_3} \tag{3.34}$$

where u_1 to u_3 are distances (mm) shown in Fig. 3.8(a).

Equations and tabulated data are given in EN 1994-1-2 for the temperatures of the profiled sheeting and of reinforcement in the slab in terms of the period of exposure, the geometry, and the density of the concrete. Reference to the data on strengths of materials at high temperatures enables the resisting tensile forces in these parts of the cross-section to be determined.

The concrete near the top of the slab is well protected from fire, so its compressive strength is assumed not to be reduced.

These assumptions enable the sagging moment of resistance, $R_{fi,d}$, to be calculated. If this does not exceed $\eta_{fi,t}R_d$ (Equation 3.33); the difference can sometimes be made up by using hogging resistance at the end of each span. The top reinforcement, provided initially to control cracking, may not be weakened by fire if it has the minimum top cover; but if it rests on the ribs of the sheeting, the compressive strength of the concrete below it will be too low for useful bending resistance to be developed.

It may be possible to neglect the concrete in the ribs, and use design data for uniform reinforced concrete slabs, from EN 1992-1-2. As an example, Fig. 3.8(b) gives the data for a uniform slab of lightweight-aggregate concrete, heated from below and not insulated, for a fire duration of 60 minutes. For concrete located x mm above the under-side of the slab, curve θ_c gives its temperature, and route ABCD gives its compressive strength, for a 'cold' cylinder strength of 25 N/mm^2.

Further explanation is given in Section 3.4.6.

3.4 Example: composite slab

The strengths of materials and characteristic actions for this structure are given in Section 3.2, and a typical floor is shown in Fig. 3.1. The following calculations illustrate the methods described in Section 3.3. In practice, the calculations may be done by the provider of the sheeting, and presented as 'safe load tables'; but here it is assumed that only the manufacturer's test data are available.

For unpropped construction, the sheeting for a span of 4 m would need to be over 100 mm deep. To reduce the floor thickness, it is assumed that the sheets will be propped at mid-span during construction. The profile chosen for trial calculations is shown in Fig. 3.9. Its overall depth is 70 mm, but the cross-section is such that the span/depth ratio based on this depth (28.6) is misleading. A more realistic value is 2000/55, which is 36.4. This may be adequate, as there is continuity over the prop between the two 2-m spans.

The next step is to choose a thickness for the composite slab, which will be designed as simply-supported over each 4-m span. The centroid of the sheeting is 30 mm above its lower surface, so the effective depth (d_p) is 120 mm and the span/depth ratio, for a slab 150 mm thick, is 4000/120, or 33.3. This is rather high for simply-supported spans, but the top re-inforcement above the supporting beams will provide some continuity. From preliminary calculations, it appears that sheeting of nominal thickness 0.9 mm may be sufficient.

It is instructive to discover why a deeper profile should have been chosen, so these initial choices are not changed.

Resistances to longitudinal shear of profiled sheetings, determined in accordance with EN 1994-1-1, are not available at the time of writing; but test data for this sheeting are sufficient to enable its shear resistance to be

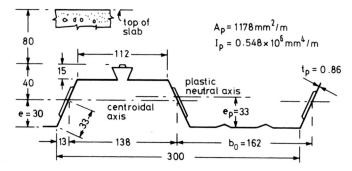

Figure 3.9 Typical cross-section of profiled sheeting and composite slab

approximated. This has been done in the *Designers' Guide to EN 1994-1-1* [17]. The data used here are given below and in Fig. 3.9.

Guaranteed minimum yield strength, $f_{yp} = 350$ N/mm^2
Design thickness, allowing for zinc coating, $t_p = 0.86$ mm
Effective area of cross-section, $A_p = 1178$ mm^2/m
Second moment of area, $I_p = 0.548 \times 10^6$ mm^4/m
Characteristic plastic moment of resistance, $M_{pa} = 6.18$ kN m/m
Distance of centroid above base, $e = 30$ mm
Distance of plastic neutral axis above base, $e_p = 33$ mm
Characteristic resistance to vertical shear, $V_{pa} = 60$ kN/m (approx.)
For resistance to longitudinal shear, $m = 184$ N/mm^2
                          $k = 0.0530$ N/mm^2
For partial-interaction design, $\tau_{u,Rd} = 0.144$ N/mm^2
Volume of concrete, 0.125 m^3 per sq. m of floor
Weight of sheeting, 0.10 kN/m^2
Weight of composite slab at 19.5 kN/m^3,
   $g_k = 0.10 + 0.125 \times 19.5 = 2.54$ kN/m^2

These data are illustrative only, and should not be relied upon in engineering practice.

3.4.1 *Profiled steel sheeting as shuttering*

In EN 1991-1-1, the density of 'unhardened concrete' is increased by 1 kN/m^3 to allow for its higher moisture content, and the imposed load during construction is 1.0 kN/m^2 (Section 3.3), so the design loads for the sheeting are:

- permanent:

$$g_d = (2.54 + 0.125) \times 1.35 = 3.60 \text{ kN/m}^2 \tag{3.35}$$

- variable:

$$q_d = 1.0 \times 1.5 = 1.5 \text{ kN/m}^2 \tag{3.36}$$

The top flanges of the supporting steel beams are assumed to be at least 150 mm wide. The bearing length for the sheeting should be at least 50 mm. Assuming that the sheeting is supported 25 mm from the flange tip (Fig. 3.10) gives the effective length of each of the two spans as

$$L_e = (4000 - 150 + 50)/2 = 1950 \text{ mm} \tag{3.37}$$

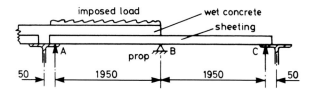

Figure 3.10 Profiled sheeting during construction

Flexure and vertical shear

The most adverse loading for sagging bending is shown in Fig. 3.10, in which the weight of the sheeting alone in span BC is neglected. Elastic analysis gives the maximum design bending moments as:

- sagging:

$$M_{Ed} = 0.0959 \times (3.6 + 1.5) \times 1.95^2 = 1.86 \text{ kN m/m}$$

- hogging (both spans loaded):

$$M_{Ed} = 0.125 \times 5.1 \times 1.95^2 = \textbf{2.42 kN m/m}$$

With $\gamma_A = 1.0$, the design resistance is $\boldsymbol{M_{Rd} = M_{pa} = \textbf{6.18 kN m/m}}$, which is ample.

Vertical shear rarely governs design of profiled sheeting. Here, the maximum value, to the left of point B in Fig. 3.10, is

$$V_{Ed} = 0.625 \times 5.1 \times 1.95 = 6.2 \text{ kN/m}$$

which is far below the design resistance of about 60 kN/m.

Deflection

The characteristic permanent load for the sheeting is $2.66 + 1.0 = 3.66$ kN/m². It is assumed that the prop does not deflect. The maximum deflection in span AB, if BC is unloaded and the sheeting is held down at C, is

$$\delta_{max} = \frac{wL_e^4}{185E_a I_p} = \frac{3.66 \times 1.95^4}{185 \times 0.21 \times 0.548} = 2.5 \text{ mm} \tag{3.38}$$

This is span/784, which is satisfactory.

3.4.2 *Composite slab – flexure and vertical shear*

This continuous slab is designed as a series of simply-supported spans. For bending, the reactions from the beams are assumed to be located as in

Fig. 3.10, so $L_e = 3.90$ m. For vertical shear, the span is taken as 4.0 m, so that the whole of the slab is included in the design loading for the beams.
From Section 3.2, the characteristic loadings are:

- $g_k = 2.54$ (slab) + 1.3 (finishes) = 3.84 kN/m^2
- $q_k = 5.0$ (imposed) + 1.2 (partitions) = 6.2 kN/m^2

The design ultimate loadings are:

- permanent:

$$g_d = 3.84 \times 1.35 = 5.18 \text{ kN/m}^2$$

- variable:

$$q_d = 6.2 \times 1.5 = 9.30 \text{ kN/m}^2$$

The mid-span bending moment is

$$M_{Ed} = 14.48 \times 3.9^2/8 = \textbf{27.6 kN m/m}$$

For the bending resistance, from Equation 3.4,

$$N_{c,f} = 1178 \times 0.35/1.0 = 412 \text{ kN/m} \tag{3.39}$$

The design compressive strength of the concrete is $0.85 \times 25/1.5 = 14.2$ N/mm^2 so, from Equation 3.5, the depth of the stress block, for full shear connection, is

$$x = 412/14.2 = 29.0 \text{ mm} \tag{3.40}$$

This is less than h_c (which can be taken as 95 mm for this profile (Fig. 3.9)), so from Equation 3.6 with $d_p = 120$ mm,

$$M_{Rd} = 412(0.12 - 0.015) = \textbf{43.3 kN m/m} \tag{3.41}$$

The bending resistance is sufficient, subject to a check on longitudinal shear.
The design vertical shear for a span of 4 m is

$$V_{Ed} = 2(5.18 + 9.3) = \textbf{29.0 kN/m}$$

For the shear resistance, from Equation 3.21 with d_p taken as 200 mm,

$$v_{min} = 0.035 \times 2^{3/2} \times 25^{1/2} = 0.49 \text{ N/mm}^2$$

From Equation 3.20 with $b_0 = 162$ mm, $b = 300$ mm (Fig. 3.9),

$$V_{Rd} = (162/300) \times 120 \times 0.49 = \textbf{31.7 kN/m} \tag{3.42}$$

which is just sufficient.

3.4.3 Composite slab – longitudinal shear

Longitudinal shear will be checked by both the 'm–k' and 'partial-interaction' methods, which are explained in Section 3.3.2. From Equation 3.19, the m–k method gives the vertical shear resistance as

$$V_{\ell,Rd} = bd_p[mA_p/(bL_s) + k]/\gamma_{Vs} = \textbf{25.9 kN/m} \tag{3.43}$$

The values used are:

$b = 1.0$ m	$m = 184$ N/mm^2
$d_p = 120$ mm	$k = 0.0530$ N/mm^2
$A_p = 1178$ mm^2/m	$\gamma_{Vs} = 1.25$
$L_s = L/4 = 1000$ mm	

where γ_{Vs} is taken from Table 1.2, and the other values are explained above.

The design vertical shear is **29.0 kN/m** (Section 3.4.2), so the slab is not strong enough, using this method.

Partial-interaction method

The mean design resistance to longitudinal shear, $\tau_{u,Rd}$, is taken as 0.144 N/mm^2 for this slab (Section 3.4). Account is taken of the shape of the profile when this value is determined from test data, so the shear resistance per metre width of sheeting is 0.144 kN per mm length. For full shear connection, the plastic neutral axis is in the slab, so the compressive force in the slab, $N_{c,f}$, is 412 kN/m, from Equation 3.39. The required length of shear span to develop this force (in the absence of any end anchorage) is

$$L_{sf} = N_{c,f}/(b\tau_{u,Rd}) = 0.412/0.144 = 2.86 \text{ m} \tag{3.44}$$

The depth of the resulting stress block in the concrete, now denoted x_f, is 29 mm, from Equation 3.40. At a distance L_x ($< L_{sf}$) from an end support, the degree of shear connection is given by

$$\eta = L_x/L_{sf} = N_c/N_{c,f} = x/x_f \tag{3.45}$$

where N_c is the force in the slab and x the depth of the stress block.

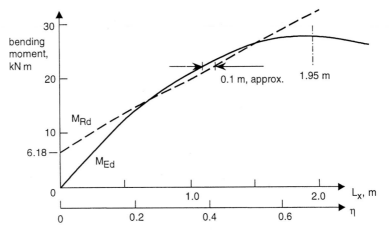

Figure 3.11   Partial-interaction method, for longitudinal shear

Equations 3.16 to 3.18 then become:

$$z = 150 - 0.5x_f - 33 + 3\eta = 102.5 + 3\eta \text{ mm}$$

$$M_{pr} = 1.25 \times 6.18(1 - \eta) \le 6.18 \text{ kN m}$$

$$M_{Rd} = 418\eta(z/1000) + M_{pr} \text{ kN m} \tag{3.46}$$

These enable M_{Rd} to be calculated for any value of η between zero and 1.0 (and hence, of L_x, from Equation 3.45). The curve so obtained is given in Fig. 3.11. The bending-moment diagram for the loading, also shown, is a parabola with maximum value 27.6 kN m/m, at mid-span (from Section 3.4.2). (A slightly more favourable curve is obtained if account is taken of the increase in lever arm, z, because the partial-interaction stress block has depth ηx_f, not x_f; but it does not give sufficient resistance here.)

It is evident that the resistance is not quite sufficient where $L_x \approx 1.0$ m ($M_{Rd} = 20.2$ kN m, $M_{Ed} = 21.0$ kN m). It would be sufficient if the curve for M_{Rd} could be moved about 0.1 m to the left. This can be achieved by the use of end anchorage. Since $\tau_{u,Rd}$ is 0.144 N/mm², an extra 'shear span' of 0.1 m would increase N_c by 14.4 kN. Assuming that one 19-mm stud connector will be welded through each trough to the supporting beam, there are 3.3 per metre, so each stud has to provide 14.4/3.3 = 4.4 kN of anchorage. It will be shown in Section 3.11.2 that the reduction in the resistance of these studs to longitudinal shear (about 40 kN) by a force of 4.4 kN in the perpendicular direction is negligible, so the studs can provide sufficient anchorage.

Another source of increased shear resistance is the friction between the sheeting and the slab above an end support. The whole of the vertical shear is assumed to be resisted by the concrete. The coefficient of friction recommended in EN 1994-1-1 is 0.5, so the additional resistance is $0.5\,V_{Ed}$, or 14.5 kN here. Coincidentally, this is precisely the force required, so end anchorage need not be relied upon.

Use of the partial-interaction method therefore enables the slab to be verified for longitudinal shear.

3.4.4 *Local effects of point load*

The design point load is $Q_{Ed} = 4.0 \times 1.5 = \mathbf{6.0\ kN}$ on any area 50 mm square, from Equation 3.3. The slab should be checked for punching shear and local bending. It is assumed that the thickness h_f of floor finish is at least 20 mm, so the data for Fig. 3.5 are:

$$b_p = a_p = 50\ \text{mm} \quad h_f = 20\ \text{mm} \quad d_p = 120\ \text{mm}$$

The small ribs at the top of the sheeting (Fig. 3.9) are neglected, so the thickness of slab above the sheeting is taken as $h_c = 95$ mm.

Punching shear
From Equation 3.22 in Section 3.3.4:

$$c_p = 2\pi h_c + 2(b_p + 2h_f) + 2(a_p + 2h_f + 2d_p - 2h_c) = 1023\ \text{mm}$$

Assuming that A193 mesh reinforcement (7-mm bars at 200-mm spacing, both ways) is provided, resting on the sheeting, the effective depths in the two directions are 76 mm and 83 mm. The reinforcement ratios are:

$$\rho_x = 0.193/76 = 0.0025 \quad \rho_y = 0.193/83 = 0.0023$$

so

$$\rho = (\rho_x \rho_y)^{1/2} = 0.0024$$

From Equation 3.23:

$$v_{Rd} = 0.12 \times 2^{1/2} \times (0.24 \times 25)^{1/3} = 0.31\ \text{N/mm}^2$$

From Equation 3.24:

$$\mathbf{V_{Rd}} = v_{Rd}c_p d = 0.31 \times 1.023 \times 79 = \mathbf{25\ kN}$$

This is a conservative value, because the ribs and sheeting have been ignored, but it far exceeds Q_{Ed}.

Local bending
From Equations 3.25 in Section 3.3.5,

$$a_m = b_m = 50 + 2(20 + 80) = 250 \text{ mm}$$

The most adverse situation for local longitudinal sagging bending is when the load is at mid-span, so L_p (Fig. 3.6) is 1.95 m. From Equation 3.26, the effective width of slab is

$$b_{em} = b_m + 2L_p[1 - (L_p/L)] = 0.25 + 3.9 \times 0.5 = 2.20 \text{ m}$$

From Equation 3.27, the sagging moment per unit width with $L_p = L/2$ is

$$m_{Ed} = Q_{Ed}L_p[1 - (L_p/L)]/b_{em} = \textbf{2.66 kN m/m}$$

which is well below the resistance of the slab, **43.3 kN m/m**.
  The transverse sagging moment under the load is given by Equation 3.28:

$$M_{Ed} = Q_{Ed}(b_{em} - b_m)/8 = 1.46 \text{ kN m}$$

This is resisted by a breadth a_m of reinforced concrete slab (Fig. 3.6), so the moment per unit width is

$$m_{Ed} = 1.46/0.25 = \textbf{5.85 kN m/m}$$

For the A193 mesh defined above, the effective depth is 76 mm and the force at yield is

$$193 \times 0.500/1.15 = 83.9 \text{ kN}$$

The depth of the concrete stress block is

$$x = 83.9/(0.85 \times 25/1.5) = 5.9 \text{ mm}$$

so the lever arm is $76 - 2.95 \simeq 73$ mm and

$$m_{Rd} = 83.9 \times 0.073 = \textbf{6.12 kN m/m}$$

which exceeds m_{Ed}.
  It will be found later that, for this slab, other limit states govern much of the slab reinforcement; but in regions where they do not, fabric of area 193 mm^2/m would probably be used, as it provides more than 0.2%, and so satisfies the empirical rule given in Section 3.3.5.

3.4.5 *Composite slab – serviceability*

Cracking of concrete above supporting beams

Following Section 3.3.6, continuity across the steel beams should be provided by reinforcement of area 0.4% of the 'area of concrete on top of the steel sheet'. For the profile of Fig. 3.9, it is not obvious whether h_c should be taken as 80 mm or 95 mm. The choice can be based on the direction of the tensile force. Here, 95 mm is used. Hence,

$$A_s = 0.004 \times 1000 \times 95 = 380 \text{ mm}^2/\text{m} \tag{3.47}$$

The detailing is best left until fire resistance has been considered.

Deflection

The characteristic load combination is used. The reinforcement of area A_s, calculated above, will provide continuity over the supports, and so reduce deflections. For this situation, EN 1994-1-1 permits the second moment of area of the slab, I, to be taken as the mean of the 'cracked' and 'uncracked' values for the section in sagging bending, and the use of a mean value of the modular ratio. This is $n = 20.2$, from Section 3.2. These values of I are 8.27 m² mm² and 13.5 m² mm², respectively, so the mean value is

$$I = 10.9 \text{ m}^2 \text{ mm}^2/\text{m} \tag{3.48}$$

The self-weight of the slab is 2.54 kN/m², so the load on the prop at B (Fig. 3.10), treated as the central support of a two-span beam, is

$$F = 2 \times 0.625 \times 2.54 \times 1.95 = 6.2 \text{ kN/m}$$

This is assumed to act as a line load on the composite slab, when the props are removed. There is in addition a load of 1.3 kN/m² from finishes (g), 1.2 kN/m² from partitions (q), and an imposed load (also q) of 5.0 kN/m².

Assuming that the props do not deflect during concreting, the mid-span deflection (for a simply-supported slab) is

$$\delta = \frac{L^3}{48EI}\left[F + \frac{5(g+q)L}{8}\right] = 3.4 + 9.9 = 13.3 \text{ mm} \tag{3.49}$$

with $L = 3.90$ m and $E = 210$ kN/mm². Hence, $\delta/L = 13.3/3900 = 1/293$.

This ratio is within the range of limits recommended in the British national annex to EN 1990 [12]. Examples are $L/300$ for removable partitions and plastered ceilings, $L/250$ for flexible floor coverings and $L/200$ for suspended ceilings.

However, the ends of the slab follow the deflections of their supporting beams. It will be found that, when these are added, the total deflection may be excessive.

3.4.6 *Composite slab – fire design*

The slab is designed for a standard fire duration of 60 minutes, using methods that are explained in Section 3.3.7, with all partial factors taken as 1.0. The effects of propping are ignored, so the characteristic loads are:

$$g_k = 3.84 \text{ kN/m}^2 \quad \text{and} \quad q_k = 6.2 \text{ kN/m}^2$$

so, from Equation 3.31,

$$\eta_{fi} = 0.56$$

For cold design, from Section 3.4.2, the mid-span bending moments are:

$$E_d = 27.6 \text{ kN m/m}$$
$$R_d = 43.3 \text{ kN m/m}$$

so, from Equation 3.32,

$$\eta_{fi,t} = 0.56(27.6/43.3) = 0.36$$

and from Equation 3.33,

$$\boldsymbol{R_{fi,d}} \geq 0.36 \times 43.3 = \textbf{15.6 kN m/m} \tag{3.50}$$

For bending resistance at mid-span, it is assumed that 8-mm reinforcing bars are located above each rib in the position shown to scale in Fig. 3.8, and also in Fig. 3.12(a). This enables them to be fixed to the A193 mesh that rests on the small top ribs shown in Fig. 3.9. Their area is 168 mm^2/m. The dimensions to the hot steel surfaces are then:

$$u_1 = 72 \text{ mm} \quad u_2 = 102 \text{ mm} \quad u_3 = 60 \text{ mm}$$

From Equation 3.34,

$$z = 0.346$$

Using formulae from Annex B of EN 1994-1-2, the temperature of this reinforcement after 60 minutes' exposure to the standard fire is found to

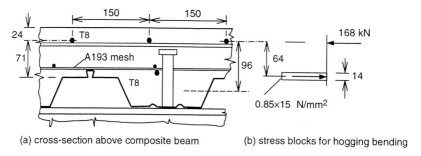

(a) cross-section above composite beam   (b) stress blocks for hogging bending

Figure 3.12   Composite slab – design for fire resistance

be 780°C, at which the design tensile stress for the steel has fallen to about 56 N/mm². The temperature of the profiled sheeting is higher, and its mean design strength is about 26 N/mm². Hence the tensile force available to resist bending is about

$$1.178 \times 26 + 0.168 \times 56 = 40 \text{ kN/m} \tag{3.51}$$

Assuming that the concrete at the top of the slab has not weakened, its stress block is about 3 mm deep, so the mean lever arm for sagging bending is about 95 mm, and

$$M_{\text{Rd,fi,sag}} = 40 \times 0.095 = \mathbf{3.8\ kN\,m/m} \tag{3.52}$$

This is less than one-quarter of the required value (Equation 3.50), so the contribution from crack-control reinforcement at the supports is now considered, using data from EN 1994-1-2 to find the hogging moment of resistance.

It is assumed that 8-mm bars at 150 mm spacing (336 mm²/m) are provided with 20 mm of top cover, as shown in Fig. 3.12(a). The yield force per unit width is

$$336 \times 0.500 = 168 \text{ kN/m}$$

As A193 mesh has been provided at the bottom of the slab, to resist local bending, the total transverse reinforcement above the beams is 336 + 193 = 529 mm²/m, which satisfies the need for 380 mm²/m for control of cracking (Section 3.4.5).

The compressive resistance of the concrete is reduced by its exposure to the fire. The concrete within the ribs is (conservatively) neglected, leaving a uniform slab of effective thickness 95 mm. The effective depth is 95 − 20 − 4 = 71 mm.

Figure 3.8(b), based on data from EN 1992-1-2 and EN 1994-1-2, shows the variation of temperature θ_c of the lightweight-aggregate concrete with its distance from the soffit of the slab, for a fire duration of 60 minutes. Following a route such as ABCD gives the design compressive strength of the concrete in fire, $f_{c,\theta}$, assuming $\gamma_C = 1.0$, because this is an accidental design situation.

For the lowest 14 mm, $f_{c,\theta} \approx 15$ N/mm², so this depth of slab can resist a compressive force $15 \times 14 \times 0.85 = 178$ kN/m. This just exceeds the yield force in the top reinforcement. Hence, the lever arm is about $71 - 7 = 64$ mm, Fig. 3.12(b), and the bending resistance is

$$M_{\text{Rd,fi,hog}} = 168 \times 0.064 = 10.7 \, \text{kN m/m} \qquad (3.53)$$

Based on a simple three-hinge plastic collapse mechanism, the total bending resistance for an internal span of floor slab is

$$M_{\text{Rd,fi}} = 3.8 + 10.7 = \textbf{14.5 kN m/m} \qquad (3.54)$$

which is still below the 15.6 kN m/m required. The deficit for an end span would be larger. Provision of more reinforcement is one remedy; others are to provide fire protection below the sheeting, or to use a deeper slab.

3.4.7 *Comments on the design of the composite slab*

It was remarked in Section 3.4 that the chosen depth of slab and size of sheeting could be inadequate for the required span and loading. The problems encountered are:

- propping was required during construction;
- the mid-span deflection of the slab may be too large when combined with the mid-span deflection of its supporting beams;
- verification for fire required extensive calculation. Verification is clearly possible for internal spans, using slightly heavier reinforcement, but there are two unsolved problems: fire resistance of an end span, and the possibility (not explored) that reliance on the top transverse reinforcement above the beams leaves the beams vulnerable to longitudinal shear failure.

3.5 Composite beams – sagging bending and vertical shear

Composite beams in buildings are usually supported by joints to steel or composite columns. The cheapest joints have little flexural strength, so it is convenient to design the beams as simply-supported. Such beams have the following advantages over beams designed as continuous at supports:

- very little of the steel web is in compression, and the steel top flange is restrained by the slab, so the resistance of the beam is not limited by buckling of steel;
- webs are less highly stressed, so it is easier to provide holes in them for the passage of services;
- bending moments and vertical shear forces are statically determinate, and are not influenced by cracking, creep, or shrinkage of concrete;
- there is no interaction between the behaviour of adjacent spans;
- bending moments in columns are lower, provided that the frame is braced against sidesway;
- no concrete at the top of the slab is in tension, except over supports;
- global analyses are simpler and design is quicker.

The disadvantages are that deflection at mid-span or crack width at supports may be excessive, and structural depth is greater than for a continuous beam.

The behaviour and design of mid-span regions of continuous beams are similar to those of simply-supported beams, considered in this chapter. The other aspects of continuous beams are treated in Chapter 4.

3.5.1 *Effective cross-section*

The presence of profiled steel sheeting in a slab is normally ignored when the slab is considered as part of the top flange of a composite beam. Longitudinal shear in the slab (explained in Section 1.6) causes shear strain in its plane, with the result that vertical cross-sections through the composite T-beam do not remain plane when it is loaded. At a cross-section, the mean longitudinal bending stress through the thickness of the slab varies across the width of the flange, as sketched in Fig. 3.13.

Simple bending theory can still give the correct value of the maximum stress (at point D) if the true flange width, B, is replaced by an effective width, b (or b_{eff}), such that the area GHJK equals the area ACDEF. Research based on elastic theory has shown that the ratio b/B depends in a complex way on the ratio of B to the span L, the type of loading, the boundary conditions at the supports, and other variables.

For simply-supported beams in buildings, EN 1994-1-1 gives the effective width as $L_e/8$ on each side of the steel web, where L_e is the distance between points of zero bending moment. The width of steel flange occupied by shear connectors, b_0, can be added, so

$$b = L_e/4 + b_0 \qquad\qquad (3.55)$$

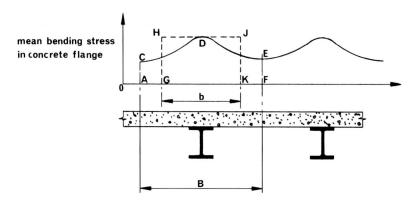

Figure 3.13   Use of effective width to allow for shear lag

provided that a width of slab $L_e/8$ is present on each side of the shear connectors.

Where profiled sheeting spans at right angles to the span of the beam (as in the worked example here), only the concrete above its ribs can resist longitudinal compression (e.g., its effective thickness in Fig. 3.9 is 80 mm). Where ribs run parallel to the span of the beam, the concrete within ribs can be included, though it is rarely necessary to do so.

Longitudinal reinforcement within the slab is usually neglected in regions of sagging bending.

3.5.2   *Classification of steel elements in compression*

Because of local buckling, the ability of a steel flange or web to resist compression depends on its slenderness, represented by its width/thickness ratio, c/t. In design to EN 1994-1-1, as in EN 1993-1-1, each flange or web in compression is placed in one of four classes. The highest (least slender) class is Class 1 (plastic). The class of a cross-section of a composite beam is the lower of the classes of its web and compression flange, and this class determines the design procedures that are available.

This well-established system is summarised in Table 3.1. The Eurocodes allow several methods of plastic global analysis, of which rigid-plastic analysis (plastic hinge analysis) is the simplest. This is considered further in Section 4.3.3.

The Eurocodes give several idealised stress–strain curves for use in plastic section analysis, of which only the simplest (rectangular stress blocks) are used in this book.

The class boundaries are defined by limiting slenderness ratios that are proportional to $(f_y)^{-0.5}$, where f_y is the nominal yield strength of the steel. This allows for the influence of yielding on loss of resistance during

Table 3.1   Classification of cross-sections, and methods of analysis

Slenderness class and name	1 Plastic	2 Compact	3 Semi-compact	4 Slender
Method of global analysis	plastic[4]	elastic	elastic	elastic
Analysis of cross-sections	plastic[4]	plastic[4]	elastic[1]	elastic[2]
Maximum ratio c/t for flanges of rolled I-sections:[3]				
uncased web	7.32	8.14	11.4	no limit
encased web	7.32	11.4	16.3	no limit

Notes: (1) hole-in-the-web method enables plastic analysis to be used;
      (2) with reduced effective width or yield strength;
      (3) for S 355 steel (f_y = 355 N/mm²);
      (4) elastic analysis may be used, but is more conservative.

buckling. The ratios in EN 1993-1-1 for steel with f_y = 355 N/mm² are given in Table 3.1 for uniformly compressed flanges of rolled I-sections with outstands of width c and thickness t. The root radius is not treated as part of the outstand.

Encasement of webs in concrete, illustrated in Fig. 3.31, is done primarily to improve resistance to fire (Section 3.10). It also prevents rotation of a flange towards the web, which occurs during local buckling, and so enables higher c/t ratios to be used at the Class 2/3 and 3/4 boundaries, as shown. At the higher compressive strains that are relied on in plastic hinge analysis, the encasement is weakened by crushing of concrete, so the c/t ratio at the Class 1/2 boundary is unchanged.

The class of a steel web is strongly influenced by the proportion of its clear depth, d, that is in compression, as shown in Fig. 3.14. For the Class 1/2 and 2/3 boundaries, plastic stress blocks are used, and the limiting d/t ratios are given in EN 1993-1-1 as functions of α, defined in Fig. 3.14. The curves show, for example, that a web with d/t = 40 moves from Class 1 to Class 3 when α increases from 0.7 to 0.8. This high rate of change is significant in the design of continuous beams (Section 4.2.1).

For the Class 3/4 boundary, elastic stress distributions are used, defined by the ratio ψ. Pure bending (no net axial force) corresponds not to α = 0.5, but to ψ = −1. In a composite T-beam in hogging bending, the elastic neutral axis is normally higher than the plastic neutral axis, and its positions for propped and unpropped construction are different, so the curve for the Class 3/4 boundary is not comparable with the others in Fig. 3.14.

For simply-supported composite beams, the steel compression flange is restrained from local buckling (and also from lateral buckling) by its connection to the concrete slab, and so is in Class 1. The plastic neutral axis for full interaction is usually within the slab or steel top flange, so the web is not in compression, when flexural failure occurs, unless partial

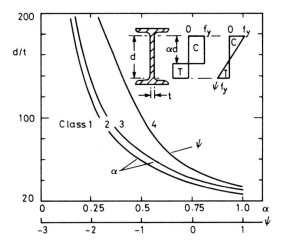

Figure 3.14 Class boundaries for webs, for $f_y = 355$ N/mm²

shear connection (Section 3.5.3.1) is used. Even then, α is sufficiently small for the web to be in Class 1 or 2. (This may not be so for the much deeper plate or box girders used in bridges.)

During construction of a composite beam, the steel beam alone may be in a lower slenderness class than the completed composite beam, and may be susceptible to lateral buckling. Design for this situation is governed by EN 1993-1-1 for steel structures.

3.5.3 *Resistance to sagging bending*

3.5.3.1 *Cross-sections in Class 1 or 2*

The methods of calculation for sections in Class 1 or 2 are in principle the same as those for composite slabs, explained in Section 3.3.1, to which reference should be made. The main assumptions are as follows:

- the tensile strength of concrete is neglected;
- plane cross-sections of the structural steel and reinforced concrete parts of a composite section each remain plane;

and, for plastic analysis of sections only:

- the effective area of the structural steel member is stressed to its design yield strength f_{yd} $(= f_y/\gamma_A)$ in tension or compression;
- the effective area of concrete in compression resists a stress of $0.85f_{cd}$ (where $f_{cd} = f_{ck}/\gamma_C$), which is constant over the whole depth between the plastic neutral axis and the most compressed fibre of the concrete.

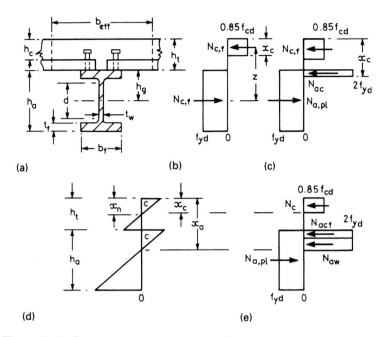

Figure 3.15 Resistance to sagging bending of composite section in Class 1 or 2

In deriving the formulae below, it is assumed that the steel member is a rolled I-section, of cross-sectional area A_a, and the slab is composite, with profiled sheeting that spans between adjacent steel members. The composite section is in Class 1 or 2, so that the whole of the design load can be assumed to be resisted by the composite member, whether the construction is propped or unpropped. This is because the inelastic behaviour that precedes flexural failure allows internal redistribution of stresses to occur.

The effective section is shown in Fig. 3.15(a). As for composite slabs, there are three common situations, as follows. The first two occur only where full shear connection is provided.

(1) Neutral axis within the concrete or composite slab
The stress blocks are shown in Fig. 3.15(b). The depth x_c, assumed to give the position of the plastic neutral axis, is found by resolving longitudinally:

$$N_{c,f} = A_a f_{yd} = b_{eff} x_c (0.85 f_{cd})$$ (3.56)

This method is valid when:

$$x_c \leq h_c$$

Taking moments about the line of action of the force in the slab,

$$M_{pl,Rd} = A_a f_{yd}(h_g + h_t - x_c/2) \tag{3.57}$$

where h_g defines the position of the centre of area of the steel section, which need not be symmetrical about its major $(y-y)$ axis.

(2) Neutral axis within the steel top flange

If Equation 3.56 gives $x_c > h_c$, then it is replaced by:

$$N_{c,f} = b_{eff} h_c (0.85 f_{cd}) \tag{3.58}$$

This force is now less than the yield force for the steel section, denoted by

$$N_{a,pl} = A_a f_{yd} \tag{3.59}$$

so the plastic neutral axis is at depth $x_c > h_t$, and is at first assumed to lie within the steel top flange (Fig. 3.15(c)). The condition for this is

$$N_{ac} = N_{a,pl} - N_{c,f} \leq 2 b t_f f_{yd} \tag{3.60}$$

The distance x_c is most easily calculated by assuming that the strength of the steel in compression is $2f_{yd}$, so that the force $N_{a,pl}$ and its line of action can be left unchanged. Resolving longitudinally to determine x_c:

$$N_{a,pl} = N_{c,f} + N_{ac} = N_{c,f} + 2b_f(x_c - h_t)f_{yd} \tag{3.61}$$

Taking moments about the line of action of the force in the slab,

$$M_{pl,Rd} = N_{a,pl}(h_g + h_t - h_c/2) - N_{ac}(x_c - h_c + h_t)/2 \tag{3.62}$$

If x_c is found to exceed $h_t + t_f$, the plastic neutral axis lies within the steel web, and $M_{pl,Rd}$ can be found by a similar method.

(3) Partial shear connection

The symbol $N_{c,f}$ was used in paragraphs (1) and (2) above for consistency with the treatment of composite slabs in Section 3.3.1. In design, its value is always the lesser of the two values given by Equations 3.56 and 3.58. It is the force that the shear connectors between the section of maximum sagging moment and each free end of the beam (a 'shear span') must be designed to resist, if full shear connection is to be provided.

Let us suppose that the shear connection is designed to resist a force N_c, smaller than $N_{c,f}$. If each connector has the same resistance to shear, and the number in each shear span is n, then the degree of shear connection is defined by:

$$\text{degree of shear connection} = \eta = n/n_f = N_c/N_{c,f} \tag{3.63}$$

where n_f is the number of connectors required for full shear connection.

The plastic moment of resistance of a composite slab with partial shear connection had to be derived in Section 3.3.1(3) by an empirical method because the flexural properties of profiled sheeting are so complex. For composite beams, simple plastic theory can be used [31].

The depth of the compressive stress block in the slab, x_c, is given by

$$x_c = N_c/(0.85f_{cd}b_{eff}) \tag{3.64}$$

and is always less than h_c. The distribution of longitudinal strain in the cross-section is intermediate between the two distributions shown (for stress) in Fig. 2.2(c), and is shown in Fig. 3.15(d), in which C means compressive strain. The neutral axis in the slab is at a depth x_n, slightly greater than x_c, as shown.

In design of reinforced concrete beams and slabs it is generally assumed that x_c/x_n is between 0.8 and 0.9. The less accurate assumption $x_c = x_n$ is made for composite beams and slabs to avoid the complexity that otherwise occurs in design when $x_c \approx h_c$ or, for beams with non-composite slabs, $x_c \approx h_t$. This introduces an error in M_{pl} that is on the unsafe side, but is negligible for composite beams. It is not negligible for composite columns, where it is allowed for (Section 5.6.5.1).

There is a second neutral axis within the steel I-section. If it lies within the steel top flange, the stress blocks are as shown in Fig. 3.15(c), except that the concrete block for the force $N_{c,f}$ is replaced by a shallower one of depth x_c, for force N_c.

Resolving longitudinally,

$$N_{ac} = N_{a,pl} - N_c$$

The depth of the neutral axis in the steel is found from

$$x_a = h_t + N_{ac}/(2b_f f_{yd}) \tag{3.65}$$

Taking moments about the line of action of N_c,

$$M_{Rd} = N_{a,pl}(h_g + h_t - x_c/2) - N_{ac}(x_a + h_t - x_c)/2 \tag{3.66}$$

If the second neutral axis lies within the steel web, the stress blocks are as shown in Fig. 3.15(e), and M_{Rd} can be found by a method similar to that for Equation 3.66.

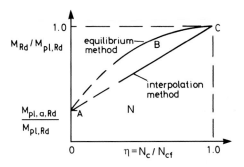

Figure 3.16 Design methods for partial shear connection

Use of partial shear connection in design

The curve ABC in Fig. 3.16 shows a typical relationship between $M_{Rd}/M_{pl,Rd}$ and degree of shear connection η, found by using the preceding equations for assumed values of η. When N_c is taken as zero, then

$$M_{Rd} = M_{pl,a,Rd}$$

where $M_{pl,a,Rd}$ is the resistance of the steel section alone.

The curve is not valid for very low degrees of shear connection, for reasons explained in Section 3.6.2. Where it is valid, it is evident that a substantial saving in the cost of shear connectors can be obtained (e.g., by using $\eta = 0.7$) when the required bending resistance M_{Ed} is only slightly below $M_{pl,Rd}$.

Where profiled sheeting is used, there is sometimes too little space in the troughs for n_f connectors to be provided within a shear span, and then partial-connection design becomes essential.

Unfortunately, curve ABC in Fig. 3.16 cannot be represented by a simple algebraic expression. In practice, it is therefore sometimes replaced (conservatively) by the line AC, given by

$$N_c = N_{c,f}(M_{Rd} - M_{pl,a,Rd})/(M_{pl,Rd} - M_{pl,a,Rd}) \tag{3.67}$$

In design, M_{Rd} is replaced by the known value M_{Ed}, and $M_{pl,Rd}$, $M_{pl,a,Rd}$ and $N_{c,f}$ are easily calculated, so this equation gives directly the design force N_c, and hence the number of connectors required in each shear span:

$$n = n_f N_c/N_{c,f} = N_c/P_{Rd} \tag{3.68}$$

where P_{Rd} is the design resistance of one connector.

The design of shear connection is further discussed in Section 3.6.

Variation in bending resistance along a span

In design, the bending resistance of a simply-supported beam is checked first at the section of maximum sagging moment, which is usually at mid-span. For a steel beam of uniform section, the bending resistance elsewhere within the span is then obviously sufficient; but this may not be so for a composite beam. Its bending resistance depends on the number of shear connectors between the nearer end support and the cross-section considered. This is shown by curve ABC in Fig. 3.16, because the x-coordinate is proportional to the number of connectors.

Suppose, for example, that a beam of span L is designed with partial shear connection and $n/n_f = 0.5$ at mid-span. Curve ABC is re-drawn in Fig. 3.17(a), with the bending resistance at mid-span, $M_{Rd,max}$, denoted by B. Length BC of this curve is not now valid, because shear failure would occur in the right-hand half span. If the connectors are uniformly spaced along the span, as is usual in buildings, then the axis n/n_f is also an axis x/L, where x is the distance from the nearer support and n is the number of connectors effective in transferring the compression to the concrete slab over a length x from a free end. Only these connectors can contribute to the bending resistance $M_{Rd,x}$ at that section, denoted E in Fig. 3.17(b). In other words, bending failure at section E would be caused (in the design model) by longitudinal shear failure along length DE of the interface between the steel flange and the concrete slab.

Which section would in fact fail first depends on the shape of the bending-moment diagram for the loading. For uniformly-distributed loading, the curve for $M_{Ed,x}$ is parabolic, and curve OFB in Fig. 3.17(a) shows that failure would occur at or near mid-span. The addition of significant point loads (e.g., from small columns) at the quarter-span points changes

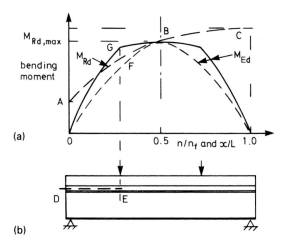

(a)

(b)

Figure 3.17   Variation of bending resistance along a span

the curve from OFB to OGB. Failure would occur near section E. A design in which M_{Ed} at mid-span was equated to $M_{Rd,max}$ would be unsafe.

This is why design codes would not allow the $0.5n_f$ connectors to be spaced uniformly along the half span, for this loading. The number required for section E would be calculated first, and spaced uniformly along DE. The remainder would be located between E and mid-span, at wider spacing. Spacing of connectors is further discussed in Section 3.6.1.

3.5.3.2 *Cross-sections in Class 3 or 4*

The resistance to bending of a beam of semi-compact or slender section is governed usually by the maximum stress in the steel section, calculated by elastic theory. Account has to be taken of the method of construction (propped or unpropped) and of the creep of concrete. The resistance may be as low as $0.7\,M_{pl,Rd}$, so it is fortunate that, in design for buildings, it is almost always possible to ensure that sections in sagging bending are in Class 1 or 2. This is more difficult for hogging bending, as explained in Section 4.2.1.

3.5.4 *Resistance to vertical shear*

In a simply-supported steel beam, bending stresses near a support are within the elastic range even when the design ultimate load is applied; but, in a composite beam, maximum slip occurs at end supports, so bending stresses cannot be found accurately by simple elastic theory based on plane sections remaining plane.

Vertical shear stresses are calculated from rates of change $(d\sigma/dx)$ of bending stresses σ, and so cannot easily be found near an end of a composite beam. It has been shown in tests that some of the vertical shear is resisted by the concrete slab, but there is no simple design model for this. The contribution from the slab is influenced by whether it is continuous across the end support, by how much it is cracked, and by local details of the shear connection.

It is therefore assumed in practice that vertical shear is resisted by the steel beam alone, exactly as if it were not composite. The web thickness of most rolled steel I-sections is sufficient to avoid buckling in shear, and then design is simple. The shear area A_v for such a section is given in EN 1993-1-1 [15] as

$$A_v = A_a - 2b_f t_f + (t_w + 2r)t_f \tag{3.69}$$

with root radius r and other notation as in Fig. 3.15(a). This shows that some of the vertical shear is resisted by the steel flanges.

The shear resistance is calculated by assuming that the yield strength in shear is $f_{yd}/\sqrt{3}$ (von Mises yield criterion), and that the whole of area A_v can reach this stress:

$$V_{pl,a,Rd} = A_v(f_{yd}/\sqrt{3}) \tag{3.70}$$

This is a 'rectangular stress block' plastic model, based essentially on test data.

The maximum slenderness of an unstiffened web for which shear buckling can be neglected is given in Eurocode 4 as

$$h_w/t_w \leq 72\varepsilon$$

where h_w is the clear distance between the flanges.

Where the steel web is encased in concrete in accordance with rules given in EN 1994-1-1, shear buckling can be neglected if

$$d/t_w \leq 124\varepsilon \tag{3.71}$$

The dimensions d and t_w are shown in Fig. 3.15(a), and

$$\varepsilon = (235/f_y)^{1/2} \tag{3.72}$$

with f_y in N/mm^2 units. This allows for the influence of yielding on shear buckling.

Interaction between bending and shear can influence the design of continuous beams, and is treated in Section 4.2.2. The large openings in webs often required for services can reduce their resistance to vertical shear. A design aid is available [32].

3.6 Composite beams – longitudinal shear

3.6.1 Critical lengths and cross-sections

As noted in Section 3.5.3.2, the bending moment at which yielding of steel first occurs in a simply-supported composite beam can be below 70% of the ultimate moment. If the bending-moment diagram is parabolic, then at ultimate load partial yielding of the steel beam can extend over half of the span.

At the interface between steel and concrete, the distribution of longitudinal shear is influenced by yielding, and also by the spacing of the connectors, their load/slip properties, and shrinkage and creep of the concrete slab. For these reasons, no attempt is made in design to calculate this distribution. Wherever possible, connectors are uniformly spaced along the span.

It was shown in Section 3.5.3 that this cannot always be done. For beams with all critical sections in Class 1 or 2, uniform spacing is allowed by EN 1994-1-1 along each *critical length*, which is a length of the interface between two adjacent *critical cross-sections*. These are:

- sections of maximum bending moment,
- supports,
- sections subjected to concentrated loads or reactions,
- places where there is a sudden change of cross-section of the beam, and
- free ends of cantilevers.

There is also a definition for tapering members.

Where the design ultimate loading is uniformly-distributed, a typical design procedure, for half the span of a beam, whether simply-supported or continuous, would be as follows.

(1) Determine the compressive force required in the concrete slab at the section of maximum sagging moment, as explained in Section 3.5.3. Let this be N_c.
(2) Determine the tensile force in the concrete slab at the support that is assumed to contribute to the bending resistance at that section (i.e., zero for a simple support, even if crack-control reinforcement is present; and the yield force in the longitudinal reinforcement, if the span is designed as continuous). Let this force be N_t.
(3) If there is a critical cross-section between these two sections, determine the force in the slab at that section. The bending moment will usually be below the yield moment, so elastic analysis of the section can be used.
(4) Choose the type of connector to be used, and determine its design resistance to shear, P_{Rd}, as explained in Section 2.5.
(5) The number of connectors required for the half span is
$$n = (N_c + N_t)/P_{Rd}. \qquad (3.73)$$

The number required within a critical length, where the change in longitudinal force is ΔN_c, is $\Delta N_c/P_{Rd}$.

An alternative to the method of step (3) would be to use the shear force diagram for the half span considered. Such a diagram is shown in Fig. 3.18 for the length ABC of a span AD, which is continuous at A and

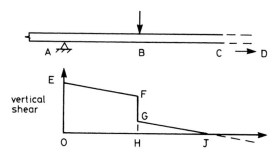

Figure 3.18 Vertical shear in a beam with an off-centre point load

has a heavy point load at B. The critical sections are A, B and C. The total number of connectors is shared between lengths AB and BC in proportion to the areas of the shear force diagram, OEFH and GJH.

In practice it might be necessary to provide a few extra connectors along BC, because codes limit the maximum spacing of connectors to prevent uplift of the slab relative to the steel beam, and to ensure that the steel top flange is sufficiently restrained from local and lateral buckling.

3.6.2 *Ductile and non-ductile connectors*

The use of uniform spacing is possible because all connectors have some ductility, or *slip capacity*. Its determination from push tests is explained in Section 2.5.

The slip capacity of headed stud connectors increases with the diameter of the shank, and has been found to be about 6 mm [33] for 19-mm studs in solid concrete slabs. Higher values have been found in tests with single studs placed centrally within the troughs of profiled steel sheeting. Off-centre studs in troughs can be much less ductile.

Slip enables longitudinal shear to be redistributed between the connectors in a critical length, before any of them fail. The slip required for this purpose increases at low degrees of shear connection, and as the critical length increases (a scale effect). A connector that is 'ductile' (has sufficient slip capacity) for a short span becomes 'non-ductile' in a long span, for which a more conservative design method must be used.

The definitions of 'ductile' connectors given in EN 1994-1-1 for headed studs welded to a steel beam with equal flanges and S355 steel are shown in Fig. 3.19. The more liberal definition given by the dashed line, for use where the slab is composite, is subject to several restrictions, based on the limited scope of current research data on slip capacity.

No design data are given in EN 1994-1-1 for shear connectors other than headed studs. If used, they should be treated as non-ductile unless their characteristic slip capacity is at least 6 mm.

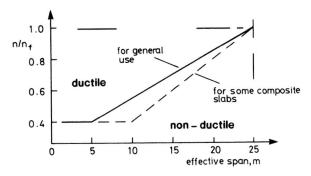

Figure 3.19 Definition of 'ductile' for welded studs, for steel sections with equal flanges

Where partial shear connection is used and the connectors are 'ductile', the bending resistance of cross-sections in Class 1 or 2 may be found by plastic theory (Equation 3.66). Otherwise, elastic theory is required, which is more complex and gives a lower resistance. Also, ductile connectors may be spaced uniformly along a critical length whereas, for non-ductile connectors, the spacing must be based on elastic analysis for longitudinal shear.

These and other rules are intended to ensure that sudden longitudinal shear failures do not occur; for example, by 'unzipping' of the shear connection, commencing from a simply-supported end of a beam. There is further explanation in Reference 17.

3.6.3 *Transverse reinforcement*

The reinforcing bars shown in Fig. 3.20 are longitudinal reinforcement for the concrete slab, to enable it to span between the beam shown and those on each side of it, but they also enhance the resistance to longitudinal shear of vertical cross-sections such as B–B. Bars provided for that purpose are known as 'transverse reinforcement', as their direction is transverse to the axis of the composite beam. Like stirrups in the web of a reinforced concrete T-beam, they supplement the shear strength of the concrete, and their behaviour can be represented by a truss analogy.

The design rules for these bars are extensive, as account has to be taken of many types and arrangements of shear connectors, of haunches, of the use of precast or composite slabs, and of interaction between the longitudinal shear stress on the section considered, v, and the transverse bending moment, shown as M_s in Fig. 3.20. The loading on the slab also causes vertical shear stress on surfaces such as B–B; but this is usually so much less than the local longitudinal shear stress, that it can be neglected.

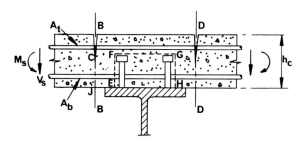

Figure 3.20   Surfaces of potential failure in longitudinal shear

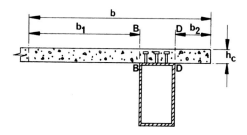

Figure 3.21   T-beam with asymmetrical concrete flange

The word 'surface' is used here because EFGH in Fig. 3.20, although not a plane, is another potential surface of shear failure. In practice, the rules for minimum height of shear connectors ensure that in slabs of uniform thickness, surfaces of type B–B are more critical; but this may not be so for haunched slabs, considered later.

The design longitudinal shear per unit length ('shear flow', denoted v_L) for surface EFGH is almost the same as that for the shear connection, and in a symmetrical T-beam half of that value is assumed to be transferred through each of the planes B–B and D–D. For an L-beam (Fig. 3.21) or where the flange of the steel beam is wide, the more accurate expressions should be used:

$$v_{L,BB} = v_L b_1/b \quad \text{and} \quad v_{L,DD} = v_L b_2/b \tag{3.74}$$

where v_L is the design shear flow for the shear connection and b is the effective width of the concrete flange.

Effective area of reinforcement

For planes such as B–B in Fig. 3.20, the effective area of transverse reinforcement per unit length of beam, A_e, is the whole of the reinforcement that is fully anchored on both sides of the plane (i.e., able to develop its yield strength in tension). This is so even where the top bars are fully stressed by the bending moment M_s, because this tension is balanced by transverse compression, which enhances the shear resistance in the region CJ by an amount at least equivalent to the contribution the reinforcement would make, in the absence of transverse bending.

3.6.3.1 *Design rules for transverse reinforcement in solid slabs*

For this subject, EN 1994-1-1 refers to EN 1992-1-1 for concrete structures. Its rules for transverse reinforcement are based on a truss analogy in which the slope of the diagonal members, θ_f, may be chosen, within defined limits, by the designer. In this explanation, the long-established angle of 45° is used.

Part of a composite beam is shown in plan in Fig. 3.22. The truss model for transverse reinforcement is illustrated by triangle ACE, in which CE represents the reinforcement for a unit length of the beam, of area A_e, and v_L is the design shear flow for a cross-section of type B–B (Fig. 3.20) above the edge of the steel flange (shown by a dashed line). The shear flow for the connectors, $2v_L$, is applied at point A, and is transferred by concrete struts AC and AE, at 45° to the axis of the beam.

The strut force is balanced at C by compression in the slab and tension in the reinforcement. The model fails when the reinforcement yields. The tensile force in it is equal to the shear on a plane such as B–B caused by the force v_L. The design equation gives the minimum area A_e of reinforcement:

$$v_{L,Ed} < v_{L,Rd} = A_e f_{sd} \tag{3.75}$$

Another requirement is that concrete struts such as AC do not fail in compression. Their design compressive stress is given in EN 1992-1-1 as

$$0.6(1 - f_{ck}/250)f_{cd} \tag{3.76}$$

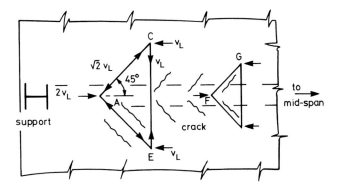

Figure 3.22   Truss model for transverse reinforcement

with f_{ck} in N/mm^2 units. The width of strut per unit length of beam is $1/\sqrt{2}$, and the force in the strut is $\sqrt{2}\,v_{L,Ed}$, so this condition is

$$v_{L,Ed} < 0.6(1 - f_{ck}/250)f_{cd}h_f/2 \qquad (3.77)$$

This rule is for normal-density concrete. For lightweight concrete, with oven-dry density ρ kg/m^3, the expression is replaced by

$$v_{L,Ed} < 0.5(0.4 + 0.6\rho/2200)(1 - f_{lck}/250)f_{lcd}h_f/2 \qquad (3.78)$$

This allows for the reduction in the ratio E_{cm}/f_{lck} as the density of the concrete reduces.

These results are assumed to be valid whatever the length of the notional struts in the slab (e.g., FG), and rely to some extent on the shear flow v_L being fairly uniform within the shear span, because the reinforcement associated with the force $2v_L$ at A is in practice provided at cross-section A, not at some point between A and mid-span. The type of cracking observed in tests where shear failure occurs, shown in Fig. 3.22, is consistent with the model.

Haunched slabs

Further design rules are required for the transverse reinforcement in haunches of the type shown in Fig. 2.1(b). These are not discussed here. Haunches encased in thin steel sheeting are considered below.

3.6.3.2 *Transverse reinforcement in composite slabs*

Where profiled sheeting spans in the direction transverse to the span of the beam, as shown in Fig. 3.15(a), it can be assumed to be effective as bottom transverse reinforcement where the sheets are continuous over the beam. Where they are not, as in the figure, the effective area of sheeting depends on how the ends of the sheets are attached to the steel top flange.

Where studs are welded to the flange through the sheeting, resistance to transverse tension is governed by local yielding of the thin sheeting around the stud. The design bearing resistance of a stud with a weld collar of diameter d_{do} in sheeting of thickness t is given in Eurocode 4 as

$$P_{pb,Rd} = k_\varphi d_{do} t f_{yp,d} \qquad (3.79)$$

where

$$k_\varphi = 1 + a/d_{do} \le 6.0 \qquad (3.80)$$

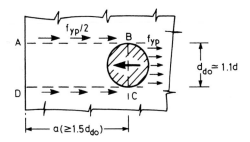

Figure 3.23 Bearing resistance of profiled sheeting, acting as transverse reinforcement

$f_{yp,d}$ is the yield strength of the sheeting, and dimension a is shown in Fig. 3.23. The formula corresponds to the assumption that yielding of the sheet occurs in direct tension along BC and in shear, at stress $f_{yp,d}/2$, along AB and CD. For pairs of studs at longitudinal spacing s (one near each edge of the steel flange), Equation 3.75 is replaced by

$$v_{L,Ed} < A_e f_{sd} + P_{pb,Rd}/s \tag{3.81}$$

The resulting reduction in the required area A_e is significant in practice where conventional studs are used; but small-diameter shot-fired pins are less effective, because of the limit $k_\varphi \leq 6$.

Where the span of the sheeting is parallel to that of the beam, transverse tension causes the corrugations to open out, so its contribution to transverse reinforcement is ignored.

3.6.4 *Detailing rules*

Where shear connectors are attached to a steel flange, there will be transverse reinforcement, and there may be a haunch (local thickening of the slab, as in Fig. 2.1(b)) or profiled steel sheeting. No reliable models exist for the three-dimensional state of stress in such a region, even in the elastic range, so the details of the design are governed by arbitrary rules of proportion, based essentially on experience.

Several of the rules given in EN 1994-1-1 are shown in Fig. 3.24. The left-hand half shows profiled sheeting that spans transversely, and the right-hand half shows a haunch.

The minimum dimensions for the head of a stud, the rule $h \geq 3d$, and the minimum projection above bottom reinforcement, are to ensure sufficient resistance to uplift. The 40-mm dimension shown is reduced to 30 mm where there is no haunch.

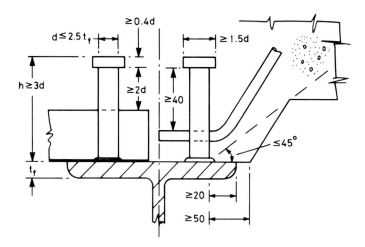

Figure 3.24 Detailing rules for shear connection

The rule $d \leq 2.5t_f$ is to avoid local failure of the steel flange, caused by load from the connector. For repeated (fatigue) loading, the limit to d/t_f is reduced to 1.5.

The 50-mm side cover to a connector and the $\leq 45°$ rule are to prevent local bursting or crushing of the concrete at the base of the connector; and the 20-mm dimension to the flange tip is to avoid local over-stress of the flange and to protect the connector from corrosion.

The minimum centre-to-centre spacing of stud connectors of diameter d is $5d$ in the longitudinal direction, $2.5d$ across the width of a steel flange in solid slabs, and $4d$ in composite slabs. These rules are to enable concrete to be properly compacted, and to avoid local overstress of the slab.

The maximum longitudinal spacing of connectors in buildings is limited to the lesser of 800 mm and six times the total slab thickness, because the transfer of shear is assumed in design to be continuous along the span, and also to avoid excessive uplift.

All such rules relevant to stresses should in principle give ratios of dimensions; where actual dimensions are given, there may be an implied assumption (e.g., that studs are between 16 mm and 22 mm in diameter), or it may be that corrosion or crack widths are relevant.

3.7 Stresses, deflections and cracking in service

A composite beam is usually designed first for ultimate limit states. Its behaviour in service must then be checked. For a simply-supported beam, the most critical serviceability limit state is usually excessive deflection, which can govern the design where unpropped construction is used. Floor structures subjected to dynamic loading (e.g., as in a dance hall or gymnasium) are also susceptible to excessive vibration (Section 3.11.3.2).

The width of cracks in concrete needs to be controlled in web-encased beams, and in hogging regions of continuous beams (Section 4.2.5).

Excessive stress in service is not itself a limit state. It may however invalidate a method of analysis (e.g., linear-elastic theory) that would otherwise be suitable for checking compliance with a serviceability criterion. No stress limits are specified in EN 1994-1-1. Where elastic analysis is used, with appropriate allowance for shear lag and creep, the policy is to modify the results, where necessary, to allow for yielding of steel and, where partial shear connection is used, for excessive slip.

If yielding of structural steel occurs in service, in a simply-supported composite beam for a building, it will be in the bottom flange, near mid-span. The likelihood of this depends on the ratio between the characteristic variable and permanent loads, given by

$$r = q_k/g_k$$

on the partial safety factors used for both actions and materials, on the method of construction used, and on the ratio of the design resistance to bending for ultimate limit states to the yield moment, which is

$$Z = M_{pl,Rd}/M_{el,Rk} \qquad (3.82)$$

where $M_{el,Rk}$ is the bending moment at which yield of steel first occurs. For sagging bending, and assuming γ_A for steel is 1.0, the ratio Z is typically between 1.25 and 1.35 for propped construction, but can rise to 1.45 or above, for unpropped construction.

Deflections are usually checked for the characteristic combination of actions, given in Equation 1.8. So, for a beam designed for distributed loads g_k and q_k only, the ratio of design bending moments (ultimate/serviceability) is

$$\mu = \frac{1.35g + 1.5q}{g + q} = \frac{1.35 + 1.5r}{1 + r} \qquad (3.83)$$

This ratio ranges from 1.42 at $r = 0.8$ to 1.45 at $r = 2.0$.

From these expressions, the stress in steel in service will reach or exceed the yield stress if $Z > \mu$. The values given above show that this is unlikely for propped construction, but could occur for unpropped construction.

Where the bending resistance of a composite section is governed by local buckling, as in a Class 3 section, elastic section analysis is used for ultimate limit states, and then stresses and/or deflections in service are less likely to influence design.

As shown below, elastic analysis of a composite section is more complex than plastic analysis, because account has to be taken of the method of construction and of the effects of creep. In principle, the following three types of loading then have to be considered separately:

- load carried by the steel beam,
- short-term load carried by the composite beam, without creep, and
- long-term load carried by the composite beam, with creep.

However, as an approximation, the composite beam may be analysed for its whole loading using a reduced value for the creep coefficient.

3.7.1   *Elastic analysis of composite sections in sagging bending*

It is assumed first that full shear connection is provided, so that the effect of slip can be neglected. All other assumptions are as for the elastic analysis of reinforced concrete sections by the method of transformed sections. The algebra is different because the flexural rigidity of the steel section alone is much greater than that of reinforcing bars.

For generality, the steel section is assumed to be asymmetrical (Fig. 3.25) with cross-sectional area A_a, second moment of area I_a, and centre of area distance z_g below the top surface of the concrete slab, which is of uniform overall thickness h_t and effective width b_{eff}.

The modular ratio for short-term loading is

$$n_0 = E_a/E_{cm}$$

where the subscript 'a' refers to structural steel and E_{cm} is the mean value of the elastic modulus for concrete, given in EN 1992-1-1. For long-term loading, a value $3n_0$ is a good approximation. For simplicity, a single value $2n_0$ is permitted for use with both types of loading. From here onwards, the symbol n is used for whatever modular ratio is appropriate, so it is defined by

$$n = E_s/E_c' \qquad (3.84)$$

where E_c' is the relevant effective modulus for the concrete. (Note: the symbol n is also used for number of shear connectors.)

It is usual to neglect reinforcement in compression, concrete in tension, and also concrete between the ribs of profiled sheeting, even when the sheeting spans longitudinally. The condition for the neutral-axis depth x to be less than h_c is

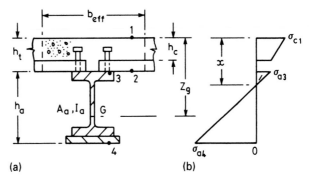

Figure 3.25 Elastic analysis of cross-section of composite beam in sagging bending

$$A_a(z_g - h_c) < b_{eff}h_c^2/(2n) \tag{3.85}$$

The neutral-axis depth is then given by the usual 'first moments of area' equation,

$$A_a(z_g - x) = b_{eff}x^2/(2n) \tag{3.86}$$

and the second moment of area, in 'steel' units, by

$$I = I_a + A_a(z_g - x)^2 + b_{eff}x^3/(3n) \tag{3.87}$$

If Condition 3.85 is not satisfied, then the neutral-axis depth exceeds h_c, as in Fig. 3.25, and is given by

$$A_a(z_g - x) = b_{eff}h_c(x - h_c/2)/n \tag{3.88}$$

The second moment of area is

$$I = I_a + A_a(z_g - x)^2 + (b_{eff}h_c/n)[h_c^2/12 + (x - h_c/2)^2] \tag{3.89}$$

In global analyses, it is sometimes convenient to use values of I based on the uncracked composite section. The values of x and I are then given by Equations 3.88 and 3.89 above, whether x exceeds h_c or not. In sagging bending, the difference between the 'cracked' and 'uncracked' values of I is usually small.

Stresses due to a sagging bending moment M are normally calculated in concrete only at level 1 in Fig. 3.25, and in steel at levels 3 and 4. These stresses are, with tensile stress positive:

$$\sigma_{c1} = -Mx/(nI) \tag{3.90}$$

$$\sigma_{a3} = M(h_t - x)/I \qquad (3.91)$$

$$\sigma_{a4} = M(h_a + h_t - x)/I \qquad (3.92)$$

Deflections

Deflections are calculated by the well-known formulae from elastic theory, using Young's modulus for structural steel. For example, the deflection of a simply-supported composite beam of span L due to distributed load q per unit length is

$$\delta_c = 5qL^4/(384E_aI) \qquad (3.93)$$

Where the shear connection is partial (i.e., $\eta < 1$, with η from Equation 3.63), the increase in deflection due to longitudinal slip depends on the method of construction. The total deflection δ is given approximately in BS 5950 [19] as

$$\delta = \delta_c[1 + k(1 - \eta)(\delta_a/\delta_c - 1)] \qquad (3.94)$$

with $k = 0.5$ for propped construction and $k = 0.3$ for unpropped construction, where δ_a is the deflection for the steel beam acting alone.

This expression is obviously correct for full shear connection ($\eta = 1$), and gives too low a result when $\eta = 0$.

EN 1994-1-1, unlike BS 5950, allows this increase in deflection to be ignored in unpropped construction where:

- either $\eta \geq 0.5$ or the forces on the connectors found by elastic analysis do not exceed $0.8\ P_{Rk}$, where P_{Rk} is their characteristic resistance, and
- for slabs with ribs transverse to the beam, the height of the ribs does not exceed 80 mm.

The arbitrary nature of these rules arises from the difficulty of predicting deflections accurately.

3.7.2   The use of limiting span-to-depth ratios

Calculations using formulae like those derived above are not only long; they are also inaccurate. It is almost as much an art as a science to predict during design the long-term deflection of a beam in a building. It is possible to allow in calculations for some of the factors that influence deflection, such as creep and shrinkage of concrete; but there are others that cannot be quantified. In developing the limiting span/depth ratios for the

British code CP 110, Beeby [34] identified nine reasons why deflections of reinforced concrete beams in service were usually less than those calculated by the designers, and increased his theoretical span/depth ratios by 36% to allow for them. Many of the reasons apply equally to composite beams, the most significant of them being the variations in the elasticity, shrinkage and creep properties of the concrete, the stiffening effect of finishes, and restraint and partial fixity at the supports.

The other problem is the difficulty of defining when a deflection becomes 'excessive'. In practice, complaints often arise from the cracking of plaster on partition walls, which can occur when the deflection of the supporting beam is as low as span/800 [34]. For partitions and in-fill panels generally, the relevant deflection is that which takes place after their construction. This can exceed that due to the finishes and imposed load, for dead-load creep deflections continue to increase for several years after construction.

It is good practice to provide partitions with appropriate joints and clearances. When this is not done the deflection under the characteristic load combination should not exceed span/300, or span/500 if the partitions are of brittle construction. Where appearance is the only criterion, a greater deflection may be acceptable where there is a suspended ceiling, and for roof beams constructed to a fall. The difficulty of assessing the accuracy and significance of a calculated deflection is such that simplified methods of calculation are justified.

3.8 Effects of shrinkage of concrete and of temperature

In the fairly dry environment of a building, an unrestrained concrete slab can be expected to shrink by 0.03% of its length (3 mm in 10 m) or more. In a composite beam, the slab is restrained by the steel member, which exerts a tensile force on it, through the shear connectors near the free ends of the beam, so its apparent shrinkage is less than the 'free' shrinkage. The forces on the shear connectors act in the opposite direction to those due to the loads, and so can be neglected in design.

The stresses due to shrinkage develop slowly, and so are reduced by creep of the concrete, but the increase they cause in the deflection of a composite beam may be significant. An approximate and usually conservative rule of thumb for estimating this deflection in a simply-supported beam is to take it as equal to the long-term deflection due to the weight of the concrete slab, excluding finishes, acting on the *composite* member.

In the beam studied in Section 3.11, this rule gives an additional deflection of 5.4 mm, whereas the calculated long-term deflection due to a shrinkage of 0.03% (with a modular ratio $n = 20.2$) is 5.9 mm.

In beams for buildings, it can usually be assumed that recommended limiting span/depth ratios are sufficiently conservative to allow for shrinkage deflections; but the designer should be alert for situations where the problem may be unusually severe (e.g., thick slabs on small steel beams, electrically heated floors, and concrete mixes with high 'free shrinkage').

In EN 1994-1-1, the effects of shrinkage need not be considered when the span/depth ratio of the beam is less than 20. For dry environments, typical values of the free shrinkage strain are given as 0.0325% for normal-weight concrete and 0.05% for lightweight concrete.

Composite beams also deflect when the slab is colder than the steel member. Such differential temperatures rarely occur in buildings, but are important in beams for bridges.

3.9 Vibration of composite floor structures

In British Standard 6472, *Guide to evaluation of human exposure to vibration in buildings* [35], the performance of a floor structure is considered to be satisfactory when the probability of annoyance to users of the floor, or of complaints from them about interference with activities, is low. There can be no simple specification of the dynamic properties that would make a floor structure 'serviceable' in this respect, because the local causes of vibration, the type of work done in the space concerned, and the psychology of its users are all relevant.

An excellent guide to this complex subject is available [36]. It and BS 6472 provided much of the basis for the following introduction to vibration design, which is limited to the situation in the design example – a typical floor of an office building, shown in Fig. 3.1.

Sources of vibration excitation
Vibration from external sources, such as highway or rail traffic, is rarely severe enough to influence design. If it is, the building should be isolated at foundation level.

Vibration from machinery in the building, such as lifts and travelling cranes, should be isolated at or near its source. In the design of a floor structure, it should be necessary to consider only sources of vibration on or near that floor. Near gymnasia or dance floors, the effects of rhythmic movement of groups of people can be troublesome; but in most buildings only two situations need be considered:

- people walking across a floor with a pace frequency of between 1.4 Hz and 2.5 Hz; and
- an impulse, such as the effect of the fall of a heavy object.

Typical reactions on floors from people walking have been analysed by Fourier series. The fundamental component has an amplitude of about 240 N. The second and third harmonics are smaller, but are relevant to design. Fundamental natural frequencies of floor structures (f_0) often lie within the frequency range of third harmonics (4.2 Hz to 7.5 Hz). The number of cycles of this harmonic, as a person walks across the span of a floor, can be sufficient for the amplitude of forced vibration to approach its steady-state value. This situation will be considered in more detail later.

Pedestrian movement causes little vibration of floor structures with f_0 exceeding about 7 Hz, but these should be checked for the effect of an impulsive load. The consequences that most influence human reactions are then the peak vertical velocity of the floor, which is proportional to the impulse, and the time for the vibration to decay, which increases with reduction in the damping ratio of the floor structure.

Human reaction to vibration
Models for human response to continuous vibration are given in BS 6472. For vibration of a floor that supports people who are standing or sitting, rather than lying down, the model consists of a base curve of root-mean-square (r.m.s.) acceleration against the fundamental natural frequency of the floor, and higher curves of similar shape. These are shown in the double logarithmic plot of Fig. 3.26. Each curve represents an approximately uniform level of human response. The base curve, denoted by $R = 1$, where R is the response factor, corresponds to a 'minimal level of adverse comment from occupants', of sensitive locations such as hospital operating theatres and precision laboratories.

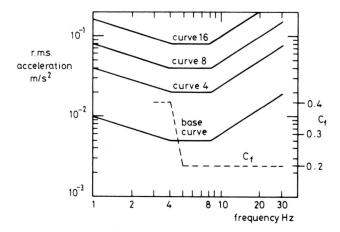

Figure 3.26 Curves of constant human response to vibration, and Fourier component factor

Curves for other values of R are obtained by multiplying the ordinates of the base curve by R. Those for $R = 4$, 8 and 16 are shown. The appropriate value of R for use in design depends on the environment. The British Standard gives:

$R = 4$ for offices,
$R = 8$ for workshops,

with the comment that use of double those values 'may result in adverse comment', which 'may increase significantly' if the magnitudes of vibration are quadrupled.

Some relaxation is possible if the vibration is not continuous. Wyatt [36] recommends that a floor subject to a person walking at resonant frequency once a minute could reasonably be permitted, a response double the value acceptable for continuous oscillation.

3.9.1 *Prediction of fundamental natural frequency*

In composite floors that need checking for vibration, damping is sufficiently low for its influence on natural frequencies to be neglected. For free elastic vibration of a beam or one-way slab of uniform section, the fundamental natural frequency is

$$f_0 = K(EI/mL^4)^{1/2} \tag{3.95}$$

where $K = \pi/2$ for simple supports and $K = 3.56$ for both ends fixed against rotation.

Values for other end conditions and multi-span members are given by Wyatt. The relevant flexural rigidity is EI (per unit width, for slabs), L is the span, and m the vibrating mass per unit length (beams) or unit area (slabs). Concrete in slabs should normally be assumed to be uncracked, and the dynamic modulus of elasticity should be used for concrete, in both beams and slabs. This modulus, E_{cd}, is typically about 8 kN/mm² higher than the static modulus, for normal-density concrete, and 3 to 6 kN/mm² higher, for lightweight-aggregate concretes of dry density not less than 1800 kg/m³. For composite beams in sagging bending, approximate allowance for these effects can be made by increasing the value of I by 10%.

Unless a more accurate estimate can be made, the mass m is usually taken as the mass of the characteristic permanent load plus 10% of the characteristic variable load.

A convenient method of calculating f_0 is to find first the mid-span deflection, δ_m say, caused by the weight of the mass m. For simply-supported members this is

$$\delta_m = 5mgL^4/(384EI)$$

Substitution for m in Equation 3.95 gives

$$f_0 = 17.8/\sqrt{\delta_m} \qquad (3.96)$$

with δ_m in millimetres.

Equation 3.96 is useful for a beam or slab considered alone. However, in a typical floor, with composite slabs continuous over a series of parallel composite beams, the total deflection (δ, say) is the sum of deflections δ_s, for the slab relative to the beams that support it, and δ_b, for the beams. A good estimate of the fundamental natural frequency is then given by

$$f_0 = 17.8/\sqrt{\delta} \qquad (3.97)$$

It follows from Equations 3.96 and 3.97 that

$$\frac{1}{f_0^2} = \frac{1}{f_{0s}^2} + \frac{1}{f_{0b}^2} \qquad (3.98)$$

where f_{0s} and f_{0b} are the frequencies for the slab and the beam, respectively, each considered alone. Equations 3.97 and 3.98 can be used also for members that are not simply-supported.

For a single-span layout of the type shown in Fig. 3.1, each beam vibrates as if simply-supported, so the length L_{eff} of the vibrating area can be taken as the span, L. The width, S, of the vibrating area will be several times the beam spacing, s. A cross-section through this area is likely to be as shown in Fig. 3.27, with most spans of the composite slab vibrating as if fixed-ended. It follows from Equation 3.95 that:

- for the beam:

$$f_{0b} = (\pi/2)(EI_b/msL^4)^{1/2} \qquad (3.99)$$

- for the slab:

$$f_{0s} = 3.56(EI_s/ms^4)^{1/2} \qquad (3.100)$$

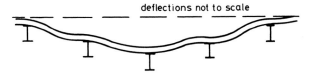

deflections not to scale

Figure 3.27   Cross-section of vibrating floor structure showing typical fundamental mode

where m is the vibrating mass per unit area, s is the spacing of the beams, and subscripts b and s mean beam and slab, respectively.

3.9.2 *Response of a composite floor to pedestrian traffic*

A method for checking the response to a single pedestrian is now given. It is assumed that the floor reaches its steady state of damped vibration under harmonic excitation from a person walking at between 1.4 Hz and 2 Hz, and that for the floor, $f_0 > 3$ Hz, to avoid resonance with the first harmonic, of typical amplitude 240 N. The effective force amplitude is

$$\bar{F} = 240C_f \tag{3.101}$$

where C_f is the Fourier component factor. It takes account of the difference between the frequency of the pedestrian's paces and the natural frequency of the floor, and is given as a function of f_0 in Fig. 3.26.

The static deflection of the floor is $\bar{F}/k_e$, where k_e is an effective stiffness. The magnification factor at resonance is $1/(2\zeta)$, where ζ is the critical damping ratio. This should normally be taken as 0.03 for open-plan offices with composite floors, though Wyatt [36] reports values as low as 0.015 for unfurnished floors. The vertical displacement y for steady-state vibration thus has frequency f_0 and is given approximately by

$$y = (\bar{F}/(2k_e\zeta)) \sin 2\pi f_0 t$$

The r.m.s. value of the acceleration is found by differentiating twice and dividing by $\sqrt{2}$:

$$a_{rms} = 4\pi^2 f_0^2 \bar{F}/(2\sqrt{2}\, k_e \zeta) \tag{3.102}$$

The effective stiffness k_e depends on the vibrating area of floor, LS. The width S can be computed in terms of the relevant flexural rigidities per unit width of floor, which are I_s and I_b/s. It is given by Wyatt as

$$S = 4.5(EI_s/mf_0^2)^{1/4} \tag{3.103}$$

This can be explained as follows. For a typical floor, f_{0b} is several times f_{0s} so, from Equation 3.98, f_{0b} is a good estimate of f_0. Substituting mf_0^2 from Equation 3.99 into Equation 3.103 gives

$$S/L = 3.6(I_s s/I_b)^{1/4}$$

Thus, the higher the ratio between the stiffnesses of the slab and the beam, the greater is the ratio of the equivalent width of the slab to the span of the beams, as would be expected.

By analogy with a simple spring–mass system, the fundamental frequency can be defined by

$$f_0 = (1/2\pi)(k_e/M_e)^{1/2} \tag{3.104}$$

where k_e is the effective stiffness. The effective mass M_e is given approximately by

$$M_e = mSL/4$$

From Equation 3.104,

$$k_e = \pi^2 f_0^2 mSL$$

With $\bar{F}$ from Equation 3.101, Equation 3.102 then gives

$$a_{rms} = 340C_f/(mSL\zeta) \tag{3.105}$$

with S given by Equation 3.103.

From the base curve in Fig. 3.26, for $4 < f_0 < 8$ Hz,

$$a_{rms} = 5 \times 10^{-3}R \text{ m/s}^2$$

so from Equation 3.105,

$$R = 68\,000C_f/(mSL\zeta) \tag{3.106}$$

in kg,m units. This equation is given on p. 28 of Reference 36.

For floors with layouts of the type shown in Fig. 3.1, and that satisfy the assumptions made above, checking for susceptibility to vibration caused by pedestrian traffic consists of finding f_0 from Equation 3.98, and a_{rms} or R, as given above, and comparing the result with the target response curve, as in Fig. 3.26.

Relevant calculations are given in Section 3.11.3.2.

The preceding summary is intended only to provide an introduction to a versatile design method, and to apply it to a single type of structure.

3.10   Fire resistance of composite beams

Fire design, based on EN 1994-1-2, *Structural fire design*, is introduced in Section 3.3.7, the whole of which is applicable to composite beams, as well as to slabs, except Section 3.3.7.5.

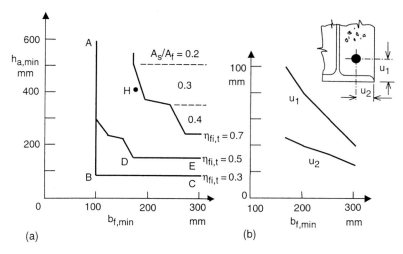

Figure 3.28   Tabulated data for web-encased beams, class R60

Beams rarely have insulation or integrity functions, and have then to be designed only for the load-bearing function, R. The fire resistance class is normally the same as that of the slab that acts as the top flange of the beam, so only the structural steel section needs further protection. This may be provided by full encasement in concrete or a lightweight fire-resisting material. A more recent method, included in EN 1994, is to encase only the web in concrete. This can be done before the beam is erected (except near end connections), and gives a cross-section of the type shown in Fig. 3.31.

In a fire, the exposed bottom flange loses its strength, but the protected web and top flange do not. For the higher load levels, given by $\eta_{fi,t}$ (defined in Section 3.3.7.2), and longer periods of fire resistance, minimum areas of longitudinal reinforcement within the encasement, A_s, are specified, in terms of the cross-sectional area, A_f, of the steel bottom flange. The minimum depth, h_a, and breadth, b_f, of the steel I-section are also specified, for each standard fire resistance period. The notation for the steel section is as in Fig. 3.15.

The requirements of EN 1994-1-2 for 60 minutes' fire exposure (class R60) are shown in Fig. 3.28. The minimum dimensions, h_a and b_f, increase with $\eta_{fi,t}$ as shown by the three sets of lines in Fig. 3.28(a). For other values of $\eta_{fi,t}$, interpolation may be used.

The minimum ratios A_s/A_f are zero for $\eta_{fi,t} = 0.3$ (ABC) and $\eta_{fi,t} = 0.5$ (ADE). For $\eta_{fi,t} = 0.7$, they are indicated within the regions where they apply. To ensure that the additional reinforcement maintains its strength for the period of fire exposure, minimum distances u_1 and u_2 are specified, in terms of the specified minimum b_f and the fire class. Those for class R60 are shown in Fig. 3.28(b).

The validity of tabulated data of this type is inevitably limited. The principal conditions for its use, given in the Eurocode, are as follows.

(a) The composite beam must be simply-supported, with

$$t_w \le b_f/15 \quad t_f \le 2t_w \quad f_y \le 355 \text{ N/mm}^2 \quad b_{eff} \le 5 \text{ m}$$

(b) The thickness of the concrete slab must be at least 120 mm, but 'thickness' is not defined. It is not clear if a value exceeding h_c in Fig. 3.15 could be used.

(c) If the slab is composite, the voids formed above the steel beam by trapezoidal profiles must be filled with fire-resistant material.

(d) The web must be encased in concrete, held in place by stirrups, fabric or stud connectors that pass through or are welded to the steel web.

The data given in Fig. 3.28 are used for the design example in Section 3.11.4. The Eurocode also gives both simple and advanced calculation models, which are often less conservative than the tabulated data, and have wider applicability. Those for beams and columns are outside the scope of this book. Other guidance is available [2].

3.11   Example: simply-supported composite beam

In this example, a typical composite T-beam is designed for the floor structure shown in Fig. 3.1, using the materials specified in Section 3.2, and the floor design given in Section 3.4. Ultimate limit states are considered first. An appropriate procedure that minimises trial and error is as follows.

(1) Choose the types and strengths of the materials to be used.

(2) Ensure that the design brief is complete. For this example it is assumed that:
   - no special provision of holes for services is required;
   - the main source of vibration is pedestrian traffic on the floor, and occupants' sensitivity to vibration is typical of that found in office buildings;
   - the specified fire resistance class is R60.

(3) Make policy decisions. For this example:
   - the steel member is to be a rolled universal beam (UB) section;
   - propped construction is to be avoided, even if this involves pre-cambering the steel beam;
   - fire resistance is to be provided by encasing the web, but not the bottom flange, in concrete;
   - nominally-pinned beam-to-column joints are to be used.

(4)   With guidance from typical span-to-depth ratios for composite beams, guess the overall depth of the beam. Assuming that the floor slab has already been designed, this gives the depth h_a of the steel section.

(5)   Guess the weight of the beam, and hence estimate the design mid-span bending moment, M_{Ed}.

(6)   Assume the lever arm to be (in the notation of Fig. 3.15)

$$(h_a/2 + h_t - h_c/2)$$

and find the required area of steel, A_a, if full shear connection is to be used, from

$$A_a f_{yd}(h_a/2 + h_t - h_c/2) \geq M_{Ed} \tag{3.107}$$

For partial shear connection, A_a should be increased.

(7)   If full shear connection is to be used, check that the yield force in the steel, $A_a f_{yd}$, is less than the compressive resistance of the concrete slab, $b_{eff} h_c(0.85 f_{cd})$. If it is not, the plastic neutral axis will be in the steel – unusual in buildings – and A_a as found above will be too small.

(8)   Knowing h_a and A_a, select a rolled steel section. Check that its web can resist the design vertical shear at an end of the beam.

(9)   Design the shear connection to provide the required bending resistance at mid-span.

(10)   Check deflections and vibration in service.

(11)   Design for fire resistance.

3.11.1   *Composite beam – full-interaction flexure and vertical shear*

From Section 3.4, the uniform characteristic loads from a 4.0-m width of floor are:

- permanent,

$$g_{k1} = 2.54 \times 4 = 10.2 \text{ kN/m} \quad \text{on steel alone}$$
$$g_{k2} = 1.3 \times 4 = 5.2 \text{ kN/m} \quad \text{on the composite beam}$$

- variable,

$$q_k = 6.2 \times 4 = 24.8 \text{ kN/m} \quad \text{on the composite beam}$$

The weight of the beam and its fire protection is estimated to be 2.2 kN/m, so the design ultimate loads are:

$$g_d = 1.35(15.4 + 2.2) = 23.7 \text{ kN/m}$$

$$q_d = 1.5 \times 24.8 \quad\quad = 37.2 \text{ kN/m}$$

The estimated depth of the steel H-section for the supporting columns is 200 mm, and the effective point of support for the beam is assumed to be 100 mm from the face of the H section, so the simply-supported span is $9.0 - 4 \times 0.10 = 8.6$ m.

The mid-span bending moment for $L = 8.6$ m is:

$$M_{\text{Ed}} = 60.9 \times 8.6^2/8 = \textbf{563 kN m} \qquad (3.108)$$

Vertical shears are required for the design of the columns and are calculated for a span of 9.0 m. It is assumed that external walls and any other load outside a 9-m width of floor are carried by edge beams that span between adjacent columns.

The design vertical shear is

$$V_{\text{Ed}} = 60.9 \times 9/2 = \textbf{274 kN} \qquad (3.109)$$

It has been assumed that the composite section will be in Class 1 or 2, so that the effects of unpropped construction can be ignored at ultimate limit states.

Deflection of this beam is likely to influence its design, because of the use of steel with $f_y = 355$ N/mm^2, lightweight-aggregate concrete and unpropped construction. The relatively low span-to-depth ratio of 16 is therefore chosen, giving an overall depth of $8600/16 = 537$ mm. The slab is 150 mm thick (Fig. 3.9) so, for the steel beam, $h_a \approx 387$ mm. From Equation 3.107 the required area of steel is

$$A_a \approx \frac{563 \times 10^6}{(355/1.0)(194 + 150 - 40)} = 5217 \text{ mm}^2$$

A suitable rolled I-section appears to be 406×140 UB 46 ($A_a = 5900$ mm^2); but its top flange may be unstable during erection, or too narrow for the profiled sheeting and shear connection. Also, with profiled sheeting it is usually necessary to use partial shear connection, so a significantly larger section is chosen, 406×178 UB 60. Its relevant properties are shown in Fig. 3.29. For its compression flange, with root radius $r = 10.2$ mm, the flange outstand is given by

$$c = (178 - 7.8)/2 - 10.2 = 74.9 \text{ mm}, \quad \text{so} \quad c/t = 74.9/12.8 = 5.9$$

which is well below the Class 1 limit of 7.32 (Table 3.1).

The weight of this beam, with normal-density web encasement at 25 kN/m^3, is

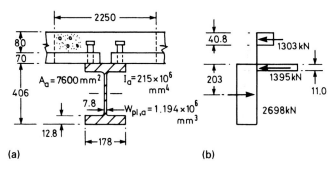

Figure 3.29 Cross-section and stress blocks for composite beam in sagging bending

$$g_k = 0.060 \times 9.81 + 25 \times (0.406 \times 0.178 - 0.0076) = 2.20 \text{ kN/m}$$

so the weight assumed earlier is correct.

For the effective flange width, it is assumed that two rows of stud connectors are 0.1 m apart, so from Equation 3.55

$$b_{\text{eff}} = 8.6/4 + 0.1 = 2.25 \text{ m} \tag{3.110}$$

Assuming that for full shear connection, the depth x of the plastic neutral axis is less than h_c (80 mm), x is found from Equation 3.56 with $f_{cd} = 25/1.5 = 16.7 \text{ N/mm}^2$:

$$N_{c,f} = 7600 \times 0.355 = 2.25x(0.85 \times 16.7)$$

whence

$$x = 84.6 \text{ mm}$$

This exceeds h_c, so the correct $N_{c,f}$ is given by Equation 3.58:

$$N_{c,f} = 2.25 \times 80(0.85 \times 16.7) = \textbf{2555 kN}$$

From Equation 3.59,

$$N_{a,pl} = A_a f_{yd} = 7600 \times 0.355 = \textbf{2698 kN} \tag{3.111}$$

Assuming that the neutral axis is in the steel top flange, Equations 3.61 give

$$N_{ac} = 2698 - 2555 = 143 \text{ kN}$$

and

$$143 = 2 \times 178(x_c - 150) \times 0.355$$

whence

$$x_c = 151.1 \text{ mm}$$

The stress blocks are as shown in Fig. 3.15(c). The distance of force N_{ac} below force N_{cf} is $151.1 - 40 = 111$ mm so, from Equation 3.62,

$$M_{pl,Rd} = 2698(0.353 - 0.04) - 143 \times 0.111 = \textbf{829 kN m} \qquad (3.112)$$

This exceeds M_{Ed} by 47%, so partial shear connection will be used, giving a concrete stress block much less than 80 mm deep.

From Equation 3.69 the shear area of this rolled section, with $r = 10.2$ mm, is

$$A_v = 7600 - 2 \times 178 \times 12.8 + 12.8(7.8 + 2 \times 10.2) = 3404 \text{ mm}^2$$

From Equation 3.70 the resistance to vertical shear is

$$V_{pl,a,Rd} = 3404(0.355/\sqrt{3}) = \textbf{697 kN} \qquad (3.113)$$

which far exceeds V_{Ed}, as is usual in composite beams for buildings when rolled steel I-sections are used.

Buckling

It is obvious that web buckling need not be considered. The steel top flange is restrained by its connection to the composite slab, and so is stable. However, its stability during erection should be checked. It is helped by the presence of web encasement. The calculation, to EN 1993-1-1, is not given here. Lateral stability of continuous beams is considered in Section 4.6.3 and in Reference 17.

3.11.2 *Composite beam – partial shear connection, and transverse reinforcement*

The minimum degree of shear connection is found next, as it may be sufficient. The condition for stud connectors to be treated as ductile when the span is 8.6 m is given by Fig. 3.19 as

$$n/n_f \geq 0.51$$

To provide an example of the use of the equilibrium method, the bending resistance is now calculated using $n = 0.51n_f$. The notation of Fig. 3.15 is used. The force N_c is 0.51 times the full-interaction value. With $N_{cf} = 2555$ kN from Section 3.11.1,

$$N_c = 0.51 \times 2555 = \mathbf{1303\ kN} \tag{3.114}$$

Since, for $N_{c,f}$,

$$x = 80\ \text{mm} \quad x_c = 0.51 \times 80 = 40.8\ \text{mm}$$

With reference to the stress blocks in Fig. 3.15(c), with $N_{a,pl} = 2698$ kN, (Equation 3.111),

$$N_{ac} = 2698 - 1303 = 1395\ \text{kN}$$

Assuming that there is a neutral axis within the steel top flange, the depth of flange in compression is

$$1395/(0.178 \times 2 \times 355) = 11.0\ \text{mm}$$

This is less than t_f (12.8 mm) so the assumption is correct, and the stress blocks are as shown in Fig. 3.29(b). Taking moments about the top surface of the slab,

$$M_{pl,Rd} = 2698 \times 0.353 - 1303 \times 0.020 - 1395 \times 0.156$$

$$= \mathbf{709\ kN\ m} \tag{3.115}$$

which exceeds M_{Ed} (563 kN m).

The interpolation method gives $M_{pl,Rd} = 630$ kN m with $n/n_f = 0.51$, so the equilibrium method is significantly less conservative. For this example, $n/n_f = 0.51$ will be used.

Number and spacing of shear connectors

It is assumed that 19-mm stud connectors will be used, 125 mm long. The length after welding is about 5 mm less, so the height of the studs is taken as 120 mm. The design shear resistance P_{Rd} is given by Equation 3.1 as 60.2 kN per stud.

For the sheeting used here, the width b_0 (Fig. 2.14) is 162 mm, from Fig. 3.9, and the other dimensions that influence the reduction factor k_t for the resistance of studs in ribs are:

$$h_p = 70\ \text{mm} \quad h = 120\ \text{mm}$$

So from Equation 2.17,

$$k_t = (0.7/\sqrt{n_r})(162/70)[(120/70) - 1] = 1.16 \quad (n_r = 1)$$

$$= 0.82 \quad (n_r \geq 2)$$

For 0.9-mm sheeting, and 19-mm studs welded through the sheeting, EN 1994-1-1 gives the following upper limits for k_t, which govern here:

for $n_r = 1$, $k_t = 0.85$, so $P_{Rd} = 51.2$ kN
for $n_r = 2$, $k_t = 0.70$, so $P_{Rd} = 42.1$ kN

From Equation 3.114 the number of single studs needed in each half span is

$$n = N_c/P_{Rd} = 1303/51.2 = 25.4, \text{ say } 26$$

For the detailing shown in Fig. 3.30, the lateral spacing of the two studs is 118 mm, or 6.2d. This exceeds the minimum given in EN 1994-1-1, which is 4d. If the separation is large, Equation 2.17 should obviously be used with $n_r = 1$, assuming that there is no interaction between the local stresses around the two studs, and giving $k_t = 0.85$. If it is small (e.g., 4d), then n_r is presumably taken as 2, so $k_t = 0.7$; and if there are two studs side-by-side on each side of the steel web, $n_r = 4$.

No upper limit for k_t is given in EN 1994-1-1 for $k_t = 4$, nor is 'large' separation defined. In a situation not covered by the code being used, as here, one has to use judgement or refer to research. Here, the separation exceeds 4d, and $k_t = 0.7$ will be used for $n_r \geq 2$.

There is one trough every 300 mm, or 16 in a half span, so it is assumed that 32 studs are provided, one in each rib on each side of the web. In Section 3.4.3, each stud was used to provide an anchorage force of 4.4 kN for the sheeting, perpendicular to the direction in which resistance is now required. The corrected values for P_{Rd} are as follows:

for $n_r = 1$, $P_{Rd}^2 = 51.2^2 - 4.4^2$ so $\boldsymbol{P_{Rd} = 51.0 \text{ kN}}$

for $n_r = 2$, $P_{Rd}^2 = 42.1^2 - 4.4^2$ so $\boldsymbol{P_{Rd} = 42.0 \text{ kN}}$ (3.116)

For 32 studs, the resistance is

$$N_{Rd} = 32 \times 42 = 1344 \text{ kN}$$

This is sufficient, so **use 1 stud in each trough on each side of the web**.

Transverse reinforcement

Rules for the use of profiled sheeting as transverse reinforcement are explained in Section 3.6.3.2. The cross-section in Fig. 3.30 illustrates compliance with the rule that the sheeting should extend at least $2d_{do}$ beyond the centre of a stud welded through it, where d_{do} is an estimate of the diameter of the stud weld, taken as 1.1d, or 20.9 mm here.

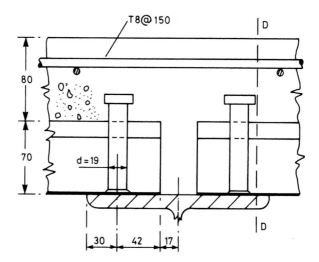

Figure 3.30 Detail of shear connection

This gives the 42-mm end distance shown. The 30-mm dimension just satisfies the relevant rule shown in Fig. 3.24. The clear gap of 34 mm between the ends of the sheeting may be reduced by tolerances, but should not fall below the minimum needed for satisfactory placing of concrete (about 25 mm).

The design longitudinal shear on a plane such as D–D in Fig. 3.30 is based on the resistance of the studs, not on the design shear flow. For $P_{Rd} = 42$ kN, the total longitudinal shear resistance is

$$v_L = 2 \times 42/0.3 = 280 \text{ kN/m}$$

The design shear for reinforcement of plane D–D is just under half this, so $v_{L,Rd}$ must be at least 140 kN/m.

The contribution from the sheeting is calculated next. Using Equations 3.79 and 3.80 with $a = 42$ mm,

$$k_\varphi = 1 + 42/20.9 = 3.0 \quad t = 0.9 \text{ mm} \quad f_{yp} = 350 \text{ N/mm}^2 \text{ and } \gamma_{Ap} = 1.0$$

so $P_{pb,Rd} = 3.0 \times 20.9 \times 0.9 \times 0.350 = 19.7$ kN

For the studs at 0.3 m spacing, $P_{pb,Rd}/s = 65.8$ kN/m. This must not exceed the tensile strength of the sheeting, which is about 412 kN/m.

Equation 3.81 is now used to find the required area of transverse reinforcement:

$$140 = 0.435A_e + 65.8 \quad \text{whence} \quad A_e = 171 \text{ mm}^2/\text{m} \tag{3.117}$$

For control of cracking of the slab above the beam, it was found (Equation 3.47) that 380 mm^2/m is required, and this governs. The detail proposed in Section 3.4.6 (8-mm bars at 150 mm spacing) plus the A193 mesh gives 529 mm^2/m. The 8-mm bars shown adjacent to the stud connector in Fig. 3.12(a) are only needed near mid-span of the slab and could, if convenient, be bent up as alternate top bars above the beams.

3.11.3   *Composite beam – deflection and vibration*

3.11.3.1   *Deflection*

The characteristic load combination for the beam is:

$$\left.\begin{array}{ll} \text{permanent (steel beam)} & g_1 = 10.2 + 2.2 = 12.4 \text{ kN/m} \\[2mm] \text{permanent (composite beam)} & g_2 = 5.2 \text{ kN/m} \\[2mm] \text{variable (composite beam)} & q = 24.8 \text{ kN/m} \end{array}\right\} \quad (3.118)$$

For a simply-supported span of 8.6 m with distributed load w kN/m and second moment of area I mm^4, the mid-span deflection is

$$\delta = \frac{5wL^4}{384EI} = \frac{5 \times 8.6^4 \times 10^9 w}{384 \times 210I} = 339 \times 10^6 \frac{w}{I} \text{ mm} \qquad (3.119)$$

For the steel beam, $I = 215 \times 10^6$ mm^4, so its deflection during construction is

$$\delta_a = 339 \times 12.4/215 = 19.5 \text{ mm (span/441)}$$

From Section 3.2, the short-term elastic modulus for the concrete is 20.7 kN/mm^2, so for variable loading the modular ratio is

$$n_0 = 210/20.7 = 10.1$$

The modular ratio for permanent load is around $3n_0$ but, for simplicity, creep will be allowed for by using $n = 2n_0$ for all loading.

The second moment of area of the composite section is calculated using Equations 3.85 to 3.89. From Fig. 3.29, relevant values are:

$$A_a = 7600 \text{ mm}^2 \quad z_g = 353 \text{ mm} \quad b_{\text{eff}} = 2250 \text{ mm} \quad I_a = 215 \times 10^6 \text{ mm}^4$$

The minimum thickness of the slab is 80 mm, but for over 90% of its area it is at least 95 mm thick (Fig. 3.9). For deflection and vibration, mean values of I are appropriate, so h_c is here taken as 95 mm.

Assuming that the neutral-axis depth exceeds h_c, Equation 3.88 gives x, now written as x_c, as

$$x_c = 176 \text{ mm} \tag{3.120}$$

From Equation 3.89,

$$10^{-6}I = 215 + 7600(0.353 - 0.176)^2 + \frac{2250 \times 95}{20.2}\left(\frac{0.095^2}{12} + 0.128^2\right)$$

$$= 215 + 240 + 181 = 636 \text{ mm}^4 \tag{3.121}$$

(Numerical values are of more convenient size in this calculation if $10^{-6}I$ is calculated, rather than I. With other values in N and mm units, bending moments are then conveniently in kN m as required, rather than N mm units.)

The deflection of the composite beam due to permanent load is

$$\delta_g = 339 \times 5.2/636 = 2.8 \text{ mm}$$

and its deflection due to variable load is

$$\delta_q = 339 \times 24.8/636 = 13.2 \text{ mm} \tag{3.122}$$

The total deflection is thus $19.5 + 2.8 + 13.2 = 35.5$ mm.

No account has yet been taken of any increase in deflection due to slip. From Section 3.7.1, EN 1994-1-1 permits it to be neglected where $n/n_f > 0.5$, which is just satisfied by the ratio 0.51 used here. However, it is instructive to calculate the maximum shear force per connector given by elastic theory, using the characteristic load applied to the composite beam. This is $5.2 + 24.8 = 30$ kN/m, so the maximum vertical shear is

$$V_{Ek} = 4.5 \times 30 = 135 \text{ kN}$$

Using the well-known result:

$$v_L = V_{Ek}(A_c/n)(x - h_c/2)/I \quad \text{(i.e., } v = VA\bar{y}/I)$$

with

$$A_c = 2250 \times 95 \text{ mm}^2 \quad n = 20.2 \quad h_c = 95 \text{ mm}$$

and x and I from Equations 3.120 and 3.121 gives the longitudinal shear flow at a support:

$$v_L = 286 \text{ kN/m}$$

The shear per connector is

$$P_{Ek} = 286 \times 0.3/2 = 43 \text{ kN}$$

From Equation 3.116, 80% of P_{Rk} is:

$$0.8P_{Rk} = 0.8 \times 1.25 \times 42.1 = 42.1 \text{ kN}$$

since $\gamma_V = 1.25$, so the alternative condition given in Section 3.7.1 is not quite satisfied.

The reader may inquire why the shear per stud for a loading of 30 kN/m, 43 kN, is almost the same as the resistance provided, 42 kN per stud (Equation 3.116), for an ultimate loading of 60.9 kN/m. The reason is that these calculations for a serviceability limit state do not allow any redistribution of force per stud along a half span. This doubles the maximum force per stud, in this case. The elastic model with full interaction and the ultimate-strength model with partial interaction happen to give similar compressive forces in the slab, for a given bending moment. The force per stud is then unaltered when the bending moment is halved.

This beam is evidently close to the borderline for deflection due to slip that underlies the rules in EN 1994-1-1. The effect of slip on deflection is now estimated, using Equation 3.94 with $k = 0.3$, $n/n_f = \eta = 0.51$, and $\delta_c = 16.0$ mm, as found above.

A load of 12.4 kN/m caused the steel beam to deflect 19.5 mm, so δ_a, for the total load on the composite beam, is

$$\delta_a = 19.5(5.2 + 24.8)/12.4 = 47.2 \text{ mm}$$

From Equation 3.94,

$$\delta = 16[1 + 0.3 \times 0.49(47.2/16 - 1)] = 20.6 \text{ mm}$$

Slip thus increases a deflection of 16 mm to over 20 mm, and the total deflection to 40 mm (span/215).

This exceeds the limit recommended in Section 3.7.2 (span/300), so the steel beam should either be propped during construction, or be cambered by an amount at least equivalent to the deflection due to permanent load, which is

$$\delta_g = 19.5 + 2.8 = 22 \text{ mm}$$

The use of 25-mm camber reduces the subsequent deflection to about 15 mm, or span/573, which should be satisfactory in most circumstances.

In practice, deflections would be slightly reduced by the stiffness of the concrete in the bottom 55 mm of the slab, and by the stiffness of the beam-to-column connections.

The use of camber was included here to illustrate the method. In practice, a better solution could be to use semi-rigid beam-to-column joints.

Maximum bending stress in the steel section

It is clear from the preceding results that yielding of the steel member under service loading is unlikely, so no allowance is needed for the effect of yielding on deflection. The maximum bending stress in the steel occurs in the bottom fibre at mid-span. It is now calculated, to illustrate the method.

Separate calculations are needed for the loadings given in Equations 3.118. For distributed load w per unit length, the stress is

$$\sigma = My/I = wL^2 y/(8I) = 9.25wy/I$$

where y is the distance of the bottom fibre below the neutral axis. From values given above, the stresses are:

for g_1: $\sigma = 9.25 \times 12.4 \times 203/215 = 108$ N/mm^2

for g_2 and q: $\sigma = 9.25 \times 30 (556 - 178)/636 = 165$ N/mm^2

The total stress is 273 N/mm^2, well below the yield stress of 355 N/mm^2. Other bending stresses can be calculated in the same way.

3.11.3.2   *Vibration*

The method given in Section 3.9 is used. It is assumed that the source of vibration is intermittent pedestrian traffic. The target value for the response factor R is 4, if the traffic is continuous, increasing to 8, if there is about one disturbance per minute.

Fundamental natural frequency

From Equations 3.118, the permanent load per beam is 17.6 kN/m. The total imposed (variable) load is 24.8 kN/m, and only one-tenth of this will be included, because vibration is likely to be worse where there are few partitions and little imposed load. The design load is thus 20.1 kN/m for beams at 4 m centres, giving a vibrating mass:

$$m = 20\ 100/(4 \times 9.81) = 512 \text{ kg/m}^2$$

For vibration, the short-term modular ratio, $n_0 = 10.1$, is used, and I is increased by 10% (Section 3.9.1). Values corresponding to those from Equations 3.120 and 3.121 are, from Table 4.5:

$$x = 129 \text{ mm} \quad 10^{-6}I_b = 1.1 \times 751 = 826 \text{ mm}^4$$

For the slab, the 'uncracked' value found in Section 3.4.5 is too low, because $n = 20.2$ was used. Similar calculations for $n_0 = 10.1$, with a 10% increase, give

$$10^{-6}I_s = 20.5 \text{ mm}^4/\text{m}$$

From Equation 3.99 with $s = 4$ m, $L = 8.6$ m,

$$f_{0b} = \frac{\pi}{2}\left(\frac{210\,000 \times 826}{512 \times 4 \times 8.6^4}\right)^{1/2} = 6.2 \text{ Hz}$$

From Equation 3.100,

$$f_{0s} = 3.56\left(\frac{210\,000 \times 20.5}{512 \times 4^4}\right)^{1/2} = 20.4 \text{ Hz}$$

From Equation 3.98,

$$f_0 = 5.9 \text{ Hz}$$

This is below 7 Hz, so no check need be made for impulsive loads.

Response of composite floor

Following Section 3.9.2, $C_f = 0.2$ from Fig. 3.26. From Equation 3.103, the vibrating width of slab is

$$S = 4.5\left(\frac{210\,000 \times 20.5}{512 \times 5.9^2}\right)^{1/4} = 17.7 \text{ m}$$

or the actual dimension of the floor normal to the span of the beams, if less. For a small value of S, the natural frequency would be higher than f_0 as calculated here, so it is assumed that the building considered is more than 17.7 m long, and this value is used.

With the critical damping ratio $\zeta = 0.03$, Equation 3.106 gives the response factor:

$$R = (68\,000 \times 0.2)/(512 \times 17.7 \times 8.6 \times 0.03) = 5.8$$

This value exceeds 4 but is well below 8. The conclusion is that continuous pedestrian traffic might result in adverse comment, but the level of movement typical of an open-plan office would not do so.

3.11.4 *Composite beam – fire design*

The method used below is explained in Sections 3.3.7 and 3.10.

From Equations 3.118 the characteristic loads per unit length of beam are:

$$g_k = 17.6 \text{ kN/m} \quad q_k = 24.8 \text{ kN/m} \quad \text{so} \quad q_k/g_k = 1.41$$

From Equation 3.31,

$$\eta_{fi} = \frac{1 + 0.7 \times 1.41}{1.35 + 1.5 \times 1.41} = 0.57$$

The beam will be designed to have a bending resistance at mid-span in fire resistance class R60. The design bending moment at mid-span for cold design is 563 kN m, from Equation 3.108 and the resistance is 709 kN m, from Equation 3.115. The resistance ratio (Equation 3.32) is

$$\eta_{fi,t} = 0.57(563/709) = 0.45$$

Fire protection is provided by encasing the web in normal-density concrete, as shown in Fig. 3.31, and by filling the voids above the steel top

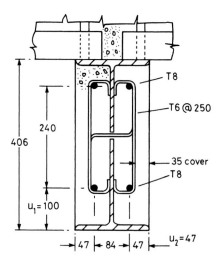

Figure 3.31 Detail of concrete-encased web

flange with fire-resistant material. This satisfies conditions (b) and (c) of Section 3.10; and all of conditions (a) are satisfied.

For the steel beam, $h_a = 406$ mm and $b_f = 178$ mm. This plots as point H in Fig. 3.28(a). As $\eta_{fi,t} < 0.5$, these dimensions qualify, and no additional reinforcement A'_s is required.

Web encasement for fire resistance is specified in EN 1994-1-2. Stirrups should be at least 6 mm in diameter, with cover not exceeding 35 mm, at spacing not exceeding 250 mm, with longitudinal bars at least 8 mm in diameter at their corners. The concrete should be connected to the steel web by one or more of:

- welding of the stirrups,
- bars passing through holes in the web,
- stud connectors.

The encasement does not increase the resistance of the composite section to sagging bending, because all of the concrete is in longitudinal tension and the additional reinforcement is neglected. The width of cracks in the concrete should be controlled, and Eurocode 4 gives rules for this purpose. Where the design crack width is 0.5 mm, it requires that the spacing of longitudinal bars should not exceed 250 mm.

A possible design for the web encasement, in accordance with these rules, is shown in Fig. 3.31. The 6-mm stirrups are either welded to the web or passed through holes in it.

In a fire, the web encasement preserves the resistance of the beam to vertical shear, and the slab is thick enough to protect the top transverse reinforcement and the shear connection.

Chapter 4
Continuous beams and slabs, and beams in frames

4.1 Introduction

A continuous beam in a building may be interrupted by its supporting internal columns; and these members are usually part of a framed structure. The type of beam-to-column joints used influences the global analysis of all the members in their vicinity. The terminology and definitions for joints used in EN 1994-1-1 follow those given in EN 1993-1-8, *Design of joints* [10], which are now summarised.

- *Connection*: location at which two members are interconnected, and assembly of connection elements and – in the case of a major axis joint – the load introduction into the column web panel.
- *Joint*: assembly of basic components that enables members to be connected together in such a way that the relevant internal forces and moments can be transferred between them.

Thus, the joint in Fig. 5.2(b), between a beam and an interior column, consists of two connections. Each connection consists of a beam end-plate, a column flange and six bolts, plus a share of the column web panel. This panel is part of both connections.

The column is part of a frame in the plane of the diagram. It could also be part of another frame, the beams of which would be attached to its web (e.g., as in Fig. 5.2(c)). This 'joint' then has two more 'connections'; but in practice this four-connection assembly is designed and detailed as two joints, one in each frame, without interaction between the two designs.

The term 'joint' is sometimes used less precisely, both here and in the Eurocodes, where the context is obvious.

Beam-to-column joints are classified both by stiffness and by strength. The categories are:

- for stiffness: rigid, semi-rigid, and nominally pinned;
- for strength: full-strength, partial-strength, and nominally pinned.

Table 4.1 Global analysis, and types of joint and joint model

Method of global analysis	Classification of joint		
Elastic	Nominally pinned	Rigid	Semi-rigid
Rigid-plastic	Nominally pinned	Full-strength	Partial-strength
Elastic-plastic	Nominally pinned	Rigid and full-strength	Semi-rigid and partial-strength Semi-rigid and full-strength Rigid and partial-strength
Type of joint model	Simple	Continuous	Semi-continuous

Table 4.2 Properties of beam-to-column joints in composite frames

Type of joint	Stiffness	Strength
Nominally pinned	Capable of transmitting the internal forces, without developing significant moments that might adversely affect members of the structure Capable of accepting the resulting rotations under the design loads	As given for Stiffness
Rigid and full strength	Has stiffness such that its deformation has no significant influence on the distribution of internal forces and moments in the structure, nor on its overall deformation Capable of transmitting the forces and moments calculated in design	Has a design resistance not less than the resistances of the members connected Has rigidity such that, under the design loads, the rotations of the necessary plastic hinges do not exceed their rotation capacities

The links between appropriate methods of global analysis, these categories, and the three types of joint model are defined in EN 1993-1-8, as shown in Table 4.1.

The most widely-used joint models in current practice are 'continuous' for bridges, and 'simple', for buildings. The scope of this book is limited to the types of joint used in the examples, which are either 'nominally pinned' or 'rigid and full strength', and to global analyses that are either elastic or rigid-plastic.

A summary of the relevant requirements for these joints, from EN 1993-1-8, is given in Table 4.2.

The use of nominally-pinned joints simplifies analysis of the structure because, at the assumed location of the pin, there is no bending moment. Action effects in beams are then independent of the properties of the columns that support them. Rigid joints transmit bending moments as well as shear forces, and the bending moments depend on the relative stiffnesses of the members joined.

If the columns of the framed structure shown in Fig. 3.1 were not encased in concrete, and the beam-to-column joints were nominal pins, the columns would be designed to EN 1993. However, the frame still satisfies the definition of 'composite frame' in EN 1994-1-1:

'*composite frame*: a framed structure in which some or all of the elements are composite members and most of the remainder are structural steel members'.

The columns shown in Fig. 3.1 are composite, and some of the top reinforcement in each floor slab may be continued into the columns, to control cracking above the ends of the beams.

In EN 1994-1-1, the definition of 'composite joint' is:

'*composite joint*: a joint between a composite member and another composite, steel, or reinforced concrete member, in which reinforcement is taken into account in design for the resistance and the stiffness of the joint'.

It follows that if the top reinforcement is 'taken into account in design', the joint is composite. It is permitted, however, to ignore light crack-control reinforcement in design for ultimate limit states, and then to design the structural steel components of the joint to EN 1993.

The example used in Chapter 5 is the two-bay nine-storey frame shown in Fig. 5.1. If nominally-pinned joints are used, the member ABC is designed as a beam with two simply-supported spans. If rigid joints are used at B, normally it is the column, not the beam, that is continuous through the joints at B, and the member ABC consists of two *beams in a frame*. The bending moments are found from analysis of the frame, or of a local region of it.

If, however, the line of columns B, E, etc., is replaced by a wall, then member ABC can be continuous at B, with no transfer of bending moment to the supporting wall. If nominally-pinned joints are used at A and C, the model for analysis is a two-span *continuous beam*, as in the example in Section 4.6, rather than beams in a frame. Such a beam does not contribute to the resistance of the frame to horizontal loading; for example, from wind. In the example in Chapter 5, this resistance is provided by reinforced concrete walls at each end of the building, to which horizontal loads are transferred by the floor slabs.

The rest of Chapter 4 refers to 'continuous beams'. In general, it applies also to 'beams in frames'. For those, the difference lies in the location and design of the joints, and in the discontinuity in the bending-moment diagram at a supporting internal column, caused by the flexural stiffness of the column.

For a given floor slab and design load per unit length of beam, the advantages of continuous beams over simple spans are:

- higher span/depth ratios can be used, for given limits to deflections;
- cracking of the top surface of a floor slab near internal columns can be controlled, so that the use of brittle finishes (e.g., terrazzo) is feasible;
- the floor structure has a higher fundamental frequency of vibration, and so is less susceptible to vibration caused by movements of people;
- the structure is more robust (e.g., in resisting the effects of fire or explosion).

The principal disadvantage is that design is more complex. Actions on one span cause action effects in adjacent spans. Even where the steel section is uniform, the stiffness and bending resistance of a composite beam vary along its length, because of cracking of concrete, changes in effective width, and variation in longitudinal reinforcement in the concrete flange.

It is not possible to predict accurately the stresses or deflections in a continuous beam, for a given set of actions. Apart from the variation over time caused by the shrinkage and creep of concrete, there are the effects of cracking of concrete. In reinforced concrete beams, these occur at cross-sections of both sagging and hogging bending, and so have little influence on distributions of bending moment. In composite beams, significant tension in concrete occurs only in hogging regions. It is influenced by the sequence of construction of the slab, the method of propping used (if any), and by effects of temperature, shrinkage and longitudinal slip.

The flexural stiffness (EI) of a fully cracked composite section can be as low as a quarter of the 'uncracked' value, so a wide variation in flexural stiffness can occur along a continuous beam of uniform section. This leads to uncertainty in the distribution of longitudinal moments, and hence in the amount of cracking to be expected. The response to a particular set of actions also depends on whether it precedes or follows another set of actions that causes cracking in a different part of the beam.

For these reasons, and also for economy, design is based as far as possible on predictions of ultimate strength (which can be checked by testing) rather than on analyses based on elastic theory. Methods have been developed from simplified models of behaviour. The limits set to the scope of some models may seem arbitrary, as they correspond to the range of available research data, rather than to known limitations of the model.

Almost the whole of Chapter 3, on simply-supported beams and slabs, applies equally to the sagging moment regions of continuous members. The properties of hogging moment regions of beams are treated in Section 4.2, which applies also to cantilevers. Then follows the global analysis of continuous beams and the calculations of stresses and deflections.

Both rolled steel I- or H-sections and small plate or box girders are considered, with or without web encasement and composite slabs. It is always assumed that the concrete slab is above the steel member, because the use of slabs below steel beams with which they are composite is rare in buildings, though it occurs in bridges. The depth of a beam can be reduced by partial embedment of the steel section within the concrete slab [37].

The use of precast or prestressed concrete floor slabs in composite frames provides an alternative to composite slabs [38]. It is outside the scope of this book.

4.2 Hogging moment regions of continuous composite beams

4.2.1 *Classification of sections, and resistance to bending*

4.2.1.1 *General*

Section 3.5.1, on effective cross-sections of beams, is applicable, except that the effective width of the concrete flange is usually less at an internal support than at mid-span. This width defines the region of the slab where longitudinal reinforcement may be assumed to contribute to the hogging moment of resistance of the beam. The plastic neutral axis always lies below the slab, so the only contribution from concrete in compression is from the web encasement, if any.

In EN 1994-1-1, the effective width in hogging bending is as explained in Section 3.5.1, except that the effective span L_e is the approximate length of the hogging moment region, which can be taken as one-quarter of each span. So at a support between spans of length L_1 and L_2, Equation 3.55 for effective width of a T-beam with pairs of stud connectors at lateral spacing b_0 becomes

$$b_{\text{eff}} = [(L_1 + L_2)/4]/4 + b_0 = (L_1 + L_2)/16 + b_0 \qquad (4.1)$$

provided that at least $b_{\text{eff}}/2$ is present on each side of the web. There is a different rule for cantilevers.

The rules for the classification of steel elements in compression (Section 3.5.2) strongly influence the design of hogging moment regions.

The proportions of rolled steel I-sections are so chosen that when they act in bending, most webs are in Class 1 or 2. But in a composite section, addition of longitudinal reinforcement in the slab rapidly increases the depth of steel web in compression, αd in Fig. 3.14. This figure shows that when $d/t > 60$, an increase in α of only 0.05 can move a web from Class 1 to Class 3, which can reduce the design moment of resistance of the section by up to 30%. This anomaly has led to a rule [3, 19] that allows a web in Class 3 to be replaced (in design) by an 'effective' web in Class 2. This 'hole-in-the-web' method is explained later. It does not apply to flanges, which can usually be designed to be in Class 1 or 2, even where plate girders are used.

Design of hogging moment regions is based on the use of full shear connection (Section 4.2.3).

4.2.1.2 *Plastic moment of resistance*

A cross-section of a composite beam in hogging bending is shown in Fig. 4.1(a). The numerical values are for the cross-section that is used in the following worked example and the diagram is to scale for these values (except for b_{eff}). The steel bottom flange is in compression, and its class is easily found, as explained in Section 3.5.2. To classify the web, the distance x_a of the plastic neutral axis above G, the centre of area of the steel section, must first be found.

Let A_s be the effective area of longitudinal reinforcement within the effective width b_{eff} of the slab. Welded mesh is normally excluded, because

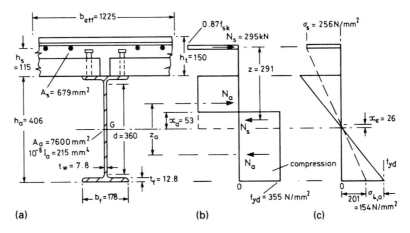

Figure 4.1 Cross-section and stress distributions for composite beam in hogging bending

it may not be sufficiently ductile to ensure that it will not fracture before the design ultimate load for the beam is reached. The design tensile force in this reinforcement is

$$N_s = A_s f_{sk}/\gamma_S = A_s f_{yd} \tag{4.2}$$

where f_{sk} is its characteristic yield strength.

If there were no tensile reinforcement, the bending resistance would be that of the steel section,

$$M_{pl,a,Rd} = W_a f_{yd} = N_a z_a \tag{4.3}$$

where W_a is the plastic section modulus and f_{yd} is the design yield strength. For rolled sections it is not necessary to calculate the forces N_a in the stress blocks of depth $h_a/2$, nor the lever arm z_a, because values of W_a are tabulated; but for plate girders N_a and z_a have to be calculated.

The simplest way of allowing for the force in the reinforcement is to assume that the stress in a depth x_a of web changes from tension to compression, where x_a is given by

$$x_a t_w (2f_{yd}) = N_s \tag{4.4}$$

provided that (as is usual)

$$x_a \leq h_a/2 - t_f$$

The depth of web in compression is given by

$$\alpha d = d/2 + x_a \tag{4.5}$$

Knowledge of α, d/t_w and f_y enables the web to be classified, as shown in Fig. 3.14 for $f_y = 355$ N/mm^2. If, by this method, a web is found to be in Class 4, the calculation should be repeated using the elastic neutral axis, as the curve that separates Class 3 from Class 4 is based on the elastic behaviour of sections. This is why, in Fig. 3.14, the ratio ψ is used, rather than α.

Concrete-encased webs in Class 3 are treated as if in Class 2 (Table 3.1), because the encasement helps to stabilise the web.

The lever arm z for the two forces N_s in Fig. 4.1(b) is given by

$$z = h_a/2 + h_s - x_a/2$$

where h_s is the height of the reinforcement above the interface. If both the compression flange and the web are in Class 1 or 2, this is the appropriate model, and the moment of resistance is

$$M_{Rd} = M_{pl.a,Rd} + N_s z \qquad (4.6)$$

If the flange is in Class 1 or 2 and the (uncased) web is in Class 3, it is still possible to use plastic section analysis, by neglecting a region in the centre of the compressed part of the web, that is assumed to be ineffective because of buckling. The calculations are more complex, as explained elsewhere [17], because this assumption changes the position of the plastic neutral axis, and in plate girders may even move it into the steel top flange. This 'hole-in-the-web' method is not available where the compression flange is in Class 3 or 4.

Example: cross-section in hogging bending

Figure 4.1(a) shows a cross-section in a region of hogging moment where the steel section is 406 × 178 UB 60 with $f_{yd} = 355$ N/mm² and dimensions as shown. Its plastic section modulus, from tables, is $W_a = 1.194 \times 10^6$ mm³. At an internal support between two spans each of 9.0 m, the longitudinal reinforcement is T12 bars, with $f_{sk} = 500$ N/mm², at 200 mm spacing. The thickness of slab above the profiled sheeting is 80 mm, so the reinforcement ratio is $36\pi/(200 \times 80) = 0.71\%$.

The top cover is 20 mm to 8-mm transverse bars (Fig. 3.12), so these bars are $20 + 8 + 12/2 = 34$ mm below the top of the slab, increased to 35 mm to allow for the ribs on both bars, giving

$$h_s = 150 - 35 = 115 \text{ mm}$$

What are the class of the section and its design resistance to hogging moments?

From Equation 4.1, assuming that $b_0 = 0.1$ m,

$$b_{eff} = (L_1 + L_2)/16 + b_0 = 18/16 + 0.1 = 1.225 \text{ m}$$

so that six T12 bars are effective, and $A_s = 679$ mm². It is assumed initially that the web is in Class 1 or 2, so that the rectangular stress blocks shown in Fig. 4.1(b) are relevant. The bottom (compression) flange has $c/t = 5.9$ (Section 3.11.1) and so is in Class 1.

From Equation 4.2 with $\gamma_S = 1.15$,

$$N_s = A_s f_{sk}/\gamma_S = 679 \times 0.500/1.15 = 295 \text{ kN}$$

From Equation 4.4,

$$x_a = N_s/(2t_w f_{yd}) = 295/(15.6 \times 0.355) = 53 \text{ mm}$$

From Equation 4.5, the ratio α is

$$\alpha = 0.5 + x_a/d = 0.5 + 53/360 = 0.647$$

The ratio d/t is 46.1. The maximum ratio in EN 1993-1-1 for a Class 2 web is $456\varepsilon/(13\alpha - 1)$ where $\varepsilon = (235/355)^{1/2} = 0.814$, so the limit is

$$d/t \leq 456 \times 0.814/(13 \times 0.647 - 1) = 50.1$$

and the web is within Class 2. This can also be seen from Fig. 3.14. From Fig. 4.1(b), the lever arm for the forces N_s is

$$z = h_a/2 + h_s - x_a/2 = 203 + 115 - 27 = 291 \text{ mm}$$

For the steel section,

$$M_{pl.a.Rd} = W_a f_{yd} = 1.194 \times 355 = 424 \text{ kN m}$$

so from Equation 4.6

$$M_{pl.Rd} = 424 + 295 \times 0.291 = \textbf{510 kN m}$$

4.2.1.3 *Elastic moment of resistance*

In the preceding calculation, it was possible to neglect the influence of the method of construction of the beam, and the effects of creep, shrinkage and temperature, because these are reduced by inelastic behaviour of the steel, and become negligible before the plastic moment of resistance is reached.

Where elastic analysis is used, creep is allowed for in the choice of the modular ratio n ($=E_a/E_{c,eff}$), and so has no influence on the properties of all-steel cross-sections. In buildings the effects of shrinkage and temperature on moments of resistance can usually be neglected, but the method of construction should be allowed for. Here, we assume that, at the section considered, the loading causes hogging bending moments $M_{a.Ed}$ in the steel member alone, and $M_{c.Ed}$ in the composite member. The small difference ($\approx 3\%$) between the elastic moduli for reinforcement and structural steel is usually neglected.

The height x_e of the elastic neutral axis of the composite section (Fig. 4.1(c)) above that of the steel section is found by taking first moments of area about the latter axis:

$$x_e(A_a + A_s) = A_s(h_a/2 + h_s) \tag{4.7}$$

and the second moment of area of the composite section is

$$I = I_a + A_a x_e^2 + A_s(h_a/2 + h_s - x_e)^2 \tag{4.8}$$

The yield moment is almost always governed by the total stress in the steel bottom flange (at level 4 in Fig. 3.25(a)). The compressive stress due to the moment $M_{a,Ed}$ is:

$$\sigma_{4,a} = M_{a,Ed}(h_a/2)/I_a \tag{4.9}$$

The remaining stress available is $f_{yd} - \sigma_{4,a}$, so the yield moment is:

$$M_{a,Ed} + M_{c,Rd} = M_{a,Ed} + \frac{(f_{yd} - \sigma_{4,a})I}{(h_a/2 + x_e)} \tag{4.10}$$

The design condition is

$$M_{c,Ed} \le M_{c,Rd} \tag{4.11}$$

The bending moment $M_{a,Ed}$ causes no stress in the slab reinforcement. In propped construction, for which $M_{a,Ed} \approx 0$, the tensile stress σ_s in these bars may govern design. It is

$$\sigma_s = M_{c,Ed}(h_a/2 + h_s - x_e)/I \tag{4.12}$$

and must not exceed f_{sd}.

Example: elastic resistance to bending

Let us now assume that the composite section shown in Fig. 4.1(a) is in Class 3, and that, at the ultimate limit state, a hogging moment of 163 kN m acts on the steel section alone, due to the use of unpropped construction.

What is the design resistance of the section to hogging moments?

From Equation 4.7 the position of the elastic neutral axis of the composite section, neglecting concrete in tension, is given by

$$x_e = \frac{679(203 + 115)}{7600 + 679} = 26 \text{ mm}$$

From Equation 4.8, the second moment of area is

$$10^{-6}I = 215 + 7600 \times 0.026^2 + 679(0.203 + 0.115 - 0.026)^2$$

$$= 278 \text{ mm}^4$$

From tables, the elastic section modulus for the steel section is $1.058 \times 10^6 \text{ mm}^3$, so the moment $M_{a,Ed}$ causes a compressive stress at level 4 (the bottom flange):

$$\sigma_{4,a} = 163/1.058 = 154 \text{ N/mm}^2$$

The design yield strength is 355 N/mm², so this leaves 201 N/mm² for resistance to the load applied to the composite member. The distance of the bottom fibre from the elastic neutral axis is $h_a/2 + x_e = 203 + 26 = 229$ mm, so the remaining resistance is

$$M_{c,Rd} = \sigma I/y = 201 \times 278/229 = 244 \text{ kN m}$$

It is evident from the stress distributions in Fig. 4.1(c) that yield occurs first in the bottom fibre. The design resistance is

$$M_{el,Rd} = M_{a,Ed} + M_{c,Rd} = 163 + 244 = \textbf{407 kN m}$$

From the preceding worked example, the shape factor is

$$S = M_{pl,Rd}/M_{el,Rd} = 510/407 = 1.25$$

4.2.2 Vertical shear, and moment-shear interaction

As explained in Section 3.5.4, vertical shear is assumed to be resisted by the web of the steel section (Equations 3.69 and 3.70). The action effect V_{Ed} must not exceed the plastic shear resistance $V_{pl,Rd}$ (or some lower value if shear buckling, not considered here, can occur).

The design rule of EN 1994-1-1 for resistance in combined bending (whether hogging or sagging) and shear is shown in Fig. 4.2. It is based on evidence from tests that there is no reduction in bending resistance until $V_{Ed} > 0.5\,V_{pl,Rd}$ (point A in the figure), and the assumption that the reduction at higher shears follows the parabolic curve AB. At point B the remaining bending resistance $M_{f,Rd}$ is that contributed by the flanges of the composite section, including the reinforcement in the slab. Along curve AB, the reduced bending resistance is given by

$$M_{v,Rd} = M_{f,Rd} + (M_{Rd} - M_{f,Rd})\left[1 - \left(\frac{2V_{Ed}}{V_{pl,Rd}} - 1\right)^2\right] \leq M_{b,Rd} \qquad (4.13)$$

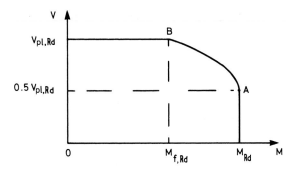

Figure 4.2 Resistance to combined bending and vertical shear

where M_{Rd} is the resistance when $V_{Ed} = 0$, and $M_{b,Rd}$ is the resistance to lateral bucking (Section 4.2.4).

When calculating $M_{f,Rd}$, it is usually accurate enough to ignore the reinforcement in the slab. When it is included, or where the steel flanges are of unequal size, only the weaker of the two flanges will be at its design yield stress.

4.2.3 *Longitudinal shear*

Section 3.6 on longitudinal shear is applicable to continuous beams and cantilevers, as well as to simply-supported spans. Some additional comments, relevant to continuous beams, are now given.

For a typical span with uniformly distributed loading, there are only three critical cross-sections: at the supports and at the section of maximum sagging moment. Points of contraflexure are not treated as critical sections because their location is different for each load case; a complication best avoided. From Equation 3.73, the number of shear connectors required for a typical critical length is

$$n = (N_c + N_t)/P_{Rd} \tag{4.14}$$

where N_t is the design tensile force in the reinforcement that is assumed to contribute to the hogging moment of resistance, and N_c is the compressive force required in the slab at mid-span, which may be less than the full-interaction value.

Full shear connection is assumed in regions of hogging moment, when n is calculated; but as the connectors may be spaced uniformly between a support and the critical section at mid-span, the number provided in the hogging region may not correspond to the force N_t.

There are several reasons for the apparently conservative requirement of EN 1994-1-1 that full shear connection be provided in hogging regions:

(1)   To compensate for some simplifications that may be unconservative:
   • neglect of the tensile strength of concrete,
   • neglect of strain-hardening of reinforcement,
   • neglect of shear due to reinforcement (e.g., welded mesh) provided for crack-width control that is neglected at ultimate limit states.
(2)   Because the design resistance of connectors, P_{Rd}, is assumed not to depend on whether the surrounding concrete is in compression or tension. There is evidence that this is slightly unconservative for hogging regions [25], but slip capacity is probably greater, which is beneficial.
(3)   For simplicity in design, including design for lateral buckling (Section 4.2.4) and for vertical shear with tension-field action [39].

The worked example in Section 4.6 illustrates the situation where the design resistance to hogging bending is that of the steel section alone, so that $N_t = 0$ in Equation 4.14, even though light reinforcement is present. It would be prudent then to provide shear connection for that reinforcement, as otherwise the uniform spacing of connectors could lead to under-provision in the sagging region.

Transverse reinforcement

As explained for regions of sagging bending, this reinforcement should be related to the shear resistance of the connectors provided, even where, for detailing reasons, their resistance exceeds the design longitudinal shear.

4.2.4   Lateral buckling

Conventional 'non-distortional' lateral torsional buckling occurs where the top flange of a simply-supported steel beam of I section has insufficient lateral restraint in the mid-span region. Both flanges are assumed to be restrained laterally at the supports, where the member may be free to rotate about a vertical axis. The top flange, in compression, is prevented by the web from buckling vertically, but if the ratio of its breadth b_f to the span L is low, it may buckle laterally as shown in Fig. 4.3(a). The cross-section rotates about a longitudinal axis, but maintains its shape.

It has to be checked that lateral-torsional buckling does not occur during casting of the concrete for an unpropped composite beam; but once the concrete has hardened, the shear connection prevents buckling of this type. The relevant design methods, being for non-composite beams, are outside the scope of this book.

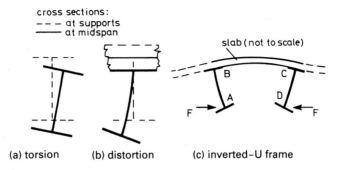

Figure 4.3   Lateral buckling

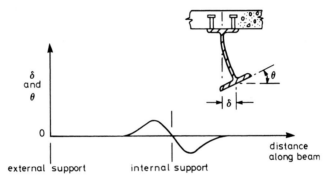

Figure 4.4   Typical deformation of steel bottom flange in distortional lateral buckling

Near internal supports of continuous beams, the compressed bottom flange of the steel section receives lateral support only through a flexible web; but the slab does prevent twisting of the steel section as a whole. The flange can only buckle if the web bends, as shown in Fig. 4.3(b). This is known as 'distortional' lateral buckling, and is the subject of this Section.

The buckle consists of a single half-wave each side of an internal support, where lateral restraint is invariably provided. The half-wave extends over most of the length of the hogging moment region. It is not sinusoidal, as the point of maximum lateral displacement is within two or three beam depths of the support, as shown in Fig. 4.4.

It is unlike local flange buckling, where the movement is essentially vertical, not lateral, and where the cross-section of maximum displacement is within one flange width of the support. There is some evidence from tests [40] that local buckling can initiate lateral buckling, but in design they are considered separately, and in different ways. Local buckling is allowed for by the classification system for steel elements in compression (Section 3.5.2). Lateral buckling is avoided by reducing the design moment of resistance at the internal support, M_{Rd}, to a lower value, $M_{b,Rd}$. Local buckling occurs where the breadth-to-thickness ratio of the flange (b_f/t_f) is high; lateral buckling occurs where it is low.

Where, as is usual in buildings, the beam is one of several parallel members, all attached to the same concrete or composite slab, design is usually based on the 'continuous inverted-U-frame' model. The tendency for the bottom flange to displace laterally causes bending of the steel web, and twisting at top-flange level, which is resisted by bending of the slab, as shown in Fig. 4.3(c).

4.2.4.1 *Elastic critical moment*

Design to EN 1994-1-1 is based on the elastic critical moment M_{cr} at the internal support. The theory for M_{cr} considers the response of a single U-frame (ABCD in Fig. 4.3(c)) to equal and opposite horizontal forces F at bottom-flange level. It leads to the following rather complex expression for M_{cr}:

$$M_{cr} = (k_c C_4/L)[(GI_{at} + k_s L^2/\pi^2)E_a I_{afz}]^{1/2} \tag{4.15}$$

where:

E_a and G are the elastic modulus and shear modulus of steel,
I_{at} is the St Venant torsion constant for the steel section,
I_{afz} is $b_f^3 t_f/12$ for the steel bottom flange, and
L is the span.

Where the steel section is symmetric about both axes, k_c is a property of the composite section (with properties A and I_y) given by

$$k_c = \frac{h_s I_y/I_{ay}}{\dfrac{h_s^2/4 + (I_{ay} + I_{az})/A_a}{e} + h_s} \tag{4.16}$$

where

$$e = \frac{A I_{ay}}{A_a z_c (A - A_a)} \tag{4.17}$$

and A_a, I_{ay} and I_{az} are properties of the structural steel section. (It should be noted that in Eurocodes, and here, subscripts y and z refer to the major

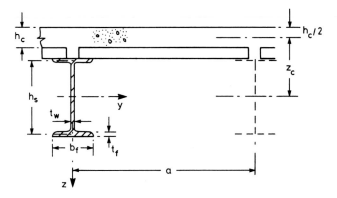

Figure 4.5 Inverted-U frame

and minor axes of a steel section, respectively. British practice has been to use x and y.) The dimensions h_s and z_c are shown in Fig. 4.5.

The term k_s is the stiffness of the U frame, per unit length along the span, that opposes lateral displacement of the bottom flange. It relates a disturbing force F per unit length of beam (Fig. 4.3(c)) to the lateral displacement of a flange, δ, caused by force F, as follows. The rotation at B that would cause displacement δ is δ/h_s; and the bending moment at B is Fh_s. The stiffness k_s is moment/rotation, so

$$k_s = Fh_s/(\delta/h_s) \quad \text{hence,} \quad \delta = Fh_s^2/k_s$$

The flexibility $1/k_s$ is the sum of the flexibilities of the slab, denoted $1/k_1$, and of the steel web, denoted $1/k_2$, so that

$$k_s = k_1 k_2/(k_1 + k_2) \tag{4.18}$$

The stiffness of the slab is represented by k_1. Where the slab is in fact continuous over the beams, even when it is designed as simply-supported, the stiffness may be taken as

$$k_1 = 4E_a I_2/a \tag{4.19}$$

where a is the spacing of the beams and $E_a I_2$ is the 'cracked' flexural stiffness of the slab above the beams.

The stiffness of the web is represented by k_2. For an uncased web,

$$k_2 = \frac{E_a t_w^3}{4(1 - v_a^2)h_s} \tag{4.20}$$

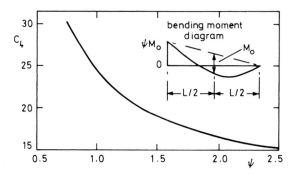

Figure 4.6   Factor C_4 for an end span of a continuous beam

where v_a is Poisson's ratio for steel. For an I-section with the web (only) encased in concrete, as used in the design example here,

$$k_2 = \frac{E_a t_w b_f^2}{16 h_s (1 + 4 n t_w / b_f)} \tag{4.21}$$

where n is the modular ratio for long-term effects, and b_f is the width of the steel flange. Equation 4.21 was derived by elastic theory, treating the concrete on one side of the web (Fig. 3.31) as a strut that restrains upwards movement of the steel bottom flange below it.

The buckling moment M_{cr} is strongly influenced by the shape of the bending-moment distribution for the span considered. This is allowed for by the factor C_4, values for which were obtained by finite-element analyses. They range from 6.2 for uniform hogging moment, to above 40 where the region of hogging moment is less than one-tenth of the span, and are given in Reference 17. Values relevant to the design example are given in Fig. 4.6.

In Equation 4.15 the term GI_{at} gives the contribution from St Venant torsion of the section. It is usually small compared with $k_s L^2 / \pi^2$ and can then be neglected with little loss of economy. The expression then becomes

$$M_{cr} \approx (k_c C_4 / \pi)(k_s E_a I_{afz})^{1/2} \tag{4.22}$$

which is independent of the span L. This enables the values of C_4 to be used for all span lengths.

Equation 4.15 for M_{cr} is valid only where rules for minimum spacing of connectors, bending stiffness of the composite slab, and proportions of the steel I-section, are satisfied. A more detailed explanation of this method and simplified versions of some of its rules are available [17].

4.2.4.2 *Buckling moment*

The value M_{cr} is relevant only to an initially perfect member that remains elastic. Evidence is limited on the influence of initial imperfections, residual stresses and yielding of steel on this type of buckling; but the Perry–Robertson formulation and the strut curves developed for overall buckling of steel columns provide a suitable basis. The method of EN 1994-1-1 is therefore as follows.

The slenderness $\bar{\lambda}_{LT}$ is given for a Class 1 or 2 section by

$$\bar{\lambda}_{LT} = (M_{Rk}/M_{cr})^{1/2} \tag{4.23}$$

where M_{Rk} is the value that would be obtained for M_{Rd} in hogging bending if the partial factors γ_A and γ_S were taken as 1.0. This is because these factors do not occur in the calculation of M_{cr}. For a Class 3 section, M_{Rk} is the characteristic yield moment.

The buckling moment is given by

$$M_{b,Rd} = \chi_{LT} M_{Rd} \tag{4.24}$$

where χ_{LT} is a function of $\bar{\lambda}_{LT}$ that in practice is taken from the relevant strut curve in Eurocode 3: Part 1.1. For rolled steel sections this curve is given by

$$\chi_{LT} = [\Phi_{LT} + (\Phi_{LT}^2 - \bar{\lambda}_{LT}^2)^{1/2}]^{-1} \quad \text{but} \quad \chi_{LT} \le 1 \tag{4.25}$$

where

$$\Phi_{LT} = 0.5[1 + \alpha_{LT}(\bar{\lambda}_{LT} - \bar{\lambda}_{LT,0}) + \beta\bar{\lambda}_{LT}^2] \tag{4.26}$$

and α_{LT} is an imperfection factor. For rolled sections, $\alpha_{LT} = 0.21$ where $h_a/b_f \le 2$ and $\alpha_{LT} = 0.34$ otherwise.

For rolled or equivalent welded steel sections, national annexes may give values for $\bar{\lambda}_{LT,0}$ that are ≤ 0.4, and for β that are ≥ 0.75. EN 1993-1-1 recommends the use of these limiting values. Their effect is that lateral buckling does not reduce M_{Rd} until $\bar{\lambda}_{LT} > 0.4$.

Simplified expression for $\bar{\lambda}_{LT}$

For cross-sections in Class 1 or 2, and with some loss of economy, Equation 4.23 can be replaced by

$$\bar{\lambda}_{LT} = 5.0\left[1 + \frac{t_w h_s}{4b_f t_f}\right]\left[\left(\frac{f_y}{E_a C_4}\right)^2\left(\frac{h_s}{t_w}\right)^3\left(\frac{t_f}{b_f}\right)\right]^{0.25}$$

provided that the steel section is symmetrical about both axes. Further details and an account of its derivation are available [17].

Exemption from check on buckling

Extensive computations based on $\bar{\lambda}_{LT} = 0.4$ have enabled conditions to be given in EN 1994-1-1 under which no detailed check on resistance to lateral buckling need be made. The principal condition relates to the overall depth h_a of the steel I-section. For IPE sections of steel with $f_y = 355$ N/mm^2,

$$h_a \le 400 \text{ mm} \tag{4.27}$$

or, if the web is encased,

$$h_a \le 600 \text{ mm} \tag{4.28}$$

The European IPE sections generally have thicker webs than British UB sections, many of which do not qualify for this relaxation. A method for determining which sections qualify is given elsewhere [17].

4.2.4.3 *Use of bracing*

Where the buckling resistance of a beam has to be checked, and has been found using Equation 4.24 to be less than the required resistance, the possibilities are as follows.

(1)   Use a steel section with a less slender web or an encased web.
(2)   Provide lateral bracing to compression flanges in the hogging moment region.

Lateral bracing is commonly used in bridges, but is less convenient in buildings, where the spacing between adjacent beams is usually wider, relative to their depth. Some examples of possible types of bracing are given in a book that covers lateral buckling of haunched composite beams [41].

Little else has been published on the use of bottom-flange bracing for beams in buildings. It interferes with the provision of services, and is best avoided.

4.2.5 *Cracking of concrete*

> '*Cracking is normal in reinforced concrete structures subject to bending, shear, torsion, or tension resulting from either direct loading or restraint of imposed deformations.*'

This clause from EN 1992-1-1 distinguishes between two types of cracking. These are treated separately in EN 1994-1-1, which follows closely the rules for crack-width control given in EN 1992. Even where no reinforcement is required to resist 'direct loading', it is necessary to limit the widths of cracks that result from tensile strains imposed on the element considered. The origin of these strains can be 'extrinsic' (external to the member), such as differential settlement of the supports of a continuous beam, or 'intrinsic' (inherent in the member), such as a temperature gradient or shrinkage of the concrete.

In reinforced concrete, cracking has little influence on tensile forces caused by direct loading, but it reduces stiffness, and so reduces the tensile force caused by an imposed deformation. For example, the tensile force in a restrained member caused by shrinkage of concrete is reduced when the member cracks.

Calculations for load-induced cracking are therefore based on the tensile force in the reinforcement after cracking (i.e., on the analysis of cracked cross-sections), whereas calculations for restraint cracking are based on the tensile force in the concrete just before it cracks.

These concepts are more difficult to apply to composite members, where there is local restraint from the axial and flexural stiffnesses of the structural steel component, applied through the shear connectors or by bond. In a web-encased beam, for example, where the steel tension flange is stressed by direct loading, the resulting strains and curvature impose a deformation on the concrete that encases the web.

The presence of structural steel members, and the differences between beam-to-column joints in composite and reinforced concrete frames, made it impossible to cover cracking in Eurocode 4 simply by cross-reference to Eurocode 2 for reinforced concrete. This led to a 'stand-alone' treatment of cracking in a slab that is part of the tension flange of a composite beam, and in the concrete encasement of a steel web.

'Cracking shall be limited to an extent that will not impair the proper functioning or durability of the structure or cause its appearance to be unacceptable.'

This quotation, also from EN 1992-1-1, refers to function and appearance. For the concrete of composite members for most types of building, the most likely cause of impairment to 'proper functioning' is corrosion of reinforcement, following breakdown of its protection by the surrounding concrete. Design is based on the exposure classes given in EN 1992-1-1. The relevant class is likely to be X0 (very dry environment) or XC1 (dry environment). For these classes, the limiting surface crack width under the quasi-permanent load combination is recommended to be 0.4 mm, with the following comment.

> '*For X0, XC1 exposure classes, crack width has no influence on durability, and this limit is set to guarantee acceptable appearance. In the absence of appearance conditions, this limit may be relaxed.*'

Tighter control of crack width is normal in bridges, and is sometimes needed in buildings: for example, in the humid environment of a laundry, or in an open-air multi-storey car park. For these and most other environments, the recommended limiting crack width is 0.3 mm for reinforced concrete, or 0.2 mm for some types of prestressed concrete. Prestressing of composite members is rare, and is not considered further. The exposure class also influences the specification for type of concrete and for minimum cover to reinforcement.

The appearance of a concrete surface may be important where a web-encased beam is visible from below, but the top surface of a slab is usually concealed by the floor finish or roof covering. Where the finish is flexible (e.g., a fitted carpet) there may be no need to specify a limit to the width of cracks; but for brittle finishes or exposed concrete surfaces, crack-width control is essential.

Limiting crack widths are normally specified as a characteristic value w_k, though EN 1992-1-1 refers also to a 'limiting calculated crack width, w_{max}', for which the same limiting values, 0.2 mm to 0.4 mm, are used. The usual interpretation of a characteristic value is one with a 5% probability of exceedence. Crack width is a random variable, but the concept of 'probability of exceedence' is difficult to apply in practice [42].

Design rules are given in EN 1994-1-1 for the following situations:

(1) where 'the control of crack width is of no interest' and beams are designed as simply-supported although the slab is continuous over supports;
(2) for the control of restraint-induced cracking, based on the tensile strength of the concrete;
(3) for load-induced cracking, with control of crack width to 0.2, 0.3 or 0.4 mm;
(4) for the calculation of estimated crack width and maximum final crack spacing.

For cases (1) to (3), simplified rules are given that do not involve the calculation of crack widths. These are outlined below. For case (4), reference is made to provisions in EN 1992-1-1 for reinforced concrete members. This situation rarely arises in buildings and is not considered further.

4.2.5.1 *No control of crack width*

'No control' is relevant to serviceability limit states. It is still necessary to ensure that the concrete retains sufficient integrity to resist shear at ultimate limit states, by acting as a continuum. It is required that the minimum longitudinal reinforcement in a concrete flange in tension shall be not less than:

- 0.4% of the area of concrete, for propped construction, or   (4.29)
- 0.2% of the area of concrete, for unpropped construction.   (4.30)

These bars are likely to yield at cracks, which may be about 0.5 mm wide, but they ensure that several cracks form rather than just one, which could be much wider. The presence of profiled steel sheeting is usually ignored, which may be conservative in some situations.

4.2.5.2 *Control of restraint-induced cracking*

Uncontrolled cracking between widely-spaced bars is avoided, and crack widths are limited, by:

- using small-diameter bars, which have better bond properties and have to be more closely spaced than larger bars;
- using 'high bond' (ribbed) bars or welded mesh;
- ensuring that the reinforcement remains elastic when cracking first occurs.

The last of these requirements is relevant to restraint cracking and leads to a design rule for minimum reinforcement, irrespective of the loading, as follows.

Let us assume that an area of concrete in uniform tension, A_{ct}, with an effective tensile strength $f_{ct,eff}$, has an area A_s of reinforcement with characteristic yield strength f_{sk}. Just before the concrete cracks, the force in it is approximately $A_{ct}f_{ct,eff}$. Cracking transfers the whole of the force to the reinforcement, which will not yield if

$$A_s f_{sk} \geq A_{ct} f_{ct,eff} \tag{4.31}$$

This condition is modified, in EN 1994-1-1, by a factor 0.8 that takes account of self-equilibrating stresses within the member (that disappear on cracking), and by a factor

$$k_c = \frac{1}{1 + (h_c/2z_0)} + 0.3 \leq 1.0 \tag{4.32}$$

that allows for the non-uniform tension in the concrete prior to cracking. In Equation 4.32, h_c is the thickness of the concrete flange, excluding any ribs, and z_0 is the distance of the centroid of the uncracked composite section (for short-term loading) below the centroid of the concrete flange. For composite beams, the ratio z_0/h_c typically increases with the span. Where it exceeds 1.17, $k_c = 1.0$, and the tension is effectively treated as uniform over the thickness of the concrete flange.

Finally, f_{sk} in Equation 4.31 is replaced by σ_s, the maximum stress permitted in the reinforcement immediately after cracking ($\leq f_{sk}$), which influences the crack width. This leads to the design rule

$$A_s \geq 0.8 k_c f_{ct,eff} A_{ct}/\sigma_s \tag{4.33}$$

To use this rule, it is necessary to estimate the value of the tensile strength $f_{ct,eff}$ when the concrete first cracks. If the intrinsic deformation due to the heat of hydration or the shrinkage of the concrete is large, cracking could occur within a week of casting, when $f_{ct,eff}$ is still low. Where this is uncertain, EN 1994-1-1 permits the use of the mean value of the tensile strength corresponding to the specified 28-day strength of the concrete, f_{ctm}, which is approximately $0.1 f_{ck}$, or $0.08 f_{cu}$, where f_{cu} is the specified cube strength.

The choice of the stress σ_s depends on the design crack width, w_k, the diameter ϕ of the reinforcing bars, and the value of $f_{ct,eff}$. Figure 4.7 gives

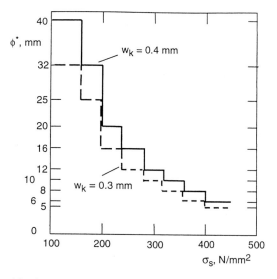

Figure 4.7 Maximum steel stress for minimum reinforcement, high bond bars

the diameters ϕ^*, for concrete with the reference strength $f_{ct,0} = 2.9$ N/mm^2. For other concrete strengths, the diameter is given in EN 1994-1-1 by

$$\phi = \phi^* f_{ct,eff} / f_{ct,0}$$

The stress σ_s may not exceed f_{sk} for the bars to be used.

4.2.5.3 *Control of load-induced cracking*

A global analysis is required, to determine the bending moment at the cross-section considered. This is usually a cross-section at an internal support, where the hogging bending moment is a maximum.

In EN 1992-1-1, the use of the quasi-permanent load combination (Section 1.3.2.3) is recommended. Imposed loads are then lower than for the characteristic combination, which is used for deciding where minimum reinforcement is required. The resulting bending-moment envelope thus has a lower proportion of each span subjected to hogging bending.

The use of the quasi-permanent combination also implies that there are no adverse effects if the cracks are wider for short periods when heavier variable loads are present. It may sometimes be necessary to check crack widths for a less probable load level, either 'frequent' or 'characteristic'. Where unpropped construction is used, load resisted by the steel member alone is excluded.

Elastic global analysis (Section 4.3.2) is used. If relative stiffnesses are based on uncracked concrete in regions where the slab is in tension, the hogging moments will be overestimated, typically by about 10%.

The tensile stress in the reinforcement nearest to the relevant concrete surface is calculated by elastic section analysis, neglecting concrete in tension. This stress, $\sigma_{s,o}$, is then increased to a value σ_s by a correction for tension stiffening, given by

$$\sigma_s = \sigma_{s,o} + 0.4 f_{ctm} A_{ct}/(\alpha_{st} A_s) \qquad (4.34)$$

where $\alpha_{st} = A I_2/A_a I_a$. The values A and I_2 are for the cracked transformed composite section, and A_a and I_a are for the structural steel section. Elastic properties of the uncracked section are also needed, to find A_{ct}, the area of concrete in tension. This is not divided by the modular ratio. A_s is the area of reinforcement within the area A_{ct}.

The correction to $\sigma_{s,o}$ is largest for lightly reinforced slabs (high A_{ct}/A_s) of strong concrete (high tensile strength, f_{ctm}). The basis of Equation 4.34 is that, until cracking is fully developed, the curvature of the steel beam equals the mean curvature of the slab, which is above-average at cracks (which increases the mean stress $\sigma_{s,o}$ to σ_s), and below-average between cracks.

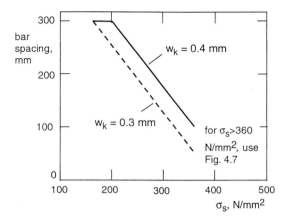

Figure 4.8 Maximum bar spacing for high bond bars

Crack control is achieved by limiting the spacing of the longitudinal reinforcing bars to the values shown in Fig. 4.8, which depend on σ_s and w_k, or by limiting the bar diameter in accordance with Fig. 4.7. It is not necessary to satisfy both requirements, for they have a common basis. This is that crack-width control relies on the bond-stress/slip property of the surface of the reinforcement, which is almost independent of bar diameter. The higher the stress σ_s, the greater is the bar perimeter required to limit the bond slip at cracks to an acceptable level.

For bars of total area A_s per unit width of slab, of diameter ϕ and at spacing s, the total bar perimeter is $u = \pi\phi/s$, and $A_s = \pi\phi^2/(4s)$. From these equations,

$$u = 4A_s/\phi = 2(\pi A_s/s)^{1/2}$$

Thus, for a given area A_s, limiting either ϕ or s will give the required value of u. The limits to ϕ and s become more severe as σ_s is increased and as w_k is reduced, as shown in Figs 4.7 and 4.8.

Fuller explanation and discussion of crack-width control for concrete flanges of composite beams is available [17, 42].

4.3 Global analysis of continuous beams

4.3.1 *General*

The subject of this Section is the determination of design values of bending moment and vertical shear for 'continuous beams' as defined in Section 4.1, caused by the actions specified for both serviceability and ultimate limit states.

Methods based on linear-elastic theory, treated in Section 4.3.2, are applicable for all limit states and all four classes of cross-section. The use of rigid-plastic analysis, also known as plastic hinge analysis, is applicable only for ultimate limit states, and is subject to the restrictions explained in Section 4.3.3. Where it can be used, the resulting members may be lighter and/or shallower, and the analyses are simpler. This is because the design moments for one span are in practice independent of the actions on adjacent spans, of variation along the span of the stiffness of the member, of the sequence and method of construction, and of the effects of temperature and of creep and shrinkage of concrete. Accurate elastic analysis has none of these advantages, so simplifications have to be made.

Section 3.5.1, on effective cross-sections, applies also to mid-span regions of continuous beams. For analysis of cross-sections, effective widths of hogging moment regions are generally narrower than those of mid-span regions (Section 4.2.1) but, for simplicity, effective widths for global analysis are assumed to be constant over the whole of each span, and are taken as the value at mid-span. This does not apply to cantilevers, where the value at the support is used.

It is assumed in global analysis that the effects of longitudinal slip are negligible. This is correct for hogging moment regions, where the use of partial shear connection is not permitted. Its use in a mid-span region slightly reduces the flexural stiffness, but for current levels of minimum shear connection, the uncertainty is probably less than that which results from cracking of concrete in hogging regions.

4.3.2 *Elastic analysis*

Elastic global analysis requires knowledge of relative (but not absolute) values of flexural stiffness (EI) over the whole length of the member analysed. Several different values of EI are required at each cross-section, as follows:

(a) for the steel member alone ($E_a I_a$), for actions applied before the member becomes composite, where unpropped construction is used;

(b) for permanent loading on the composite member ($E_a I$), where I is determined, in 'steel' units, by the method of transformed sections using a modular ratio, $n = E_a/E_c'$, where E_c' is an effective modulus that allows for creep of concrete;

(c) for variable loading on the composite member, as above, except that the modular ratio is $n_0 = E_a/E_{cm}$, and E_{cm} is the mean secant modulus for short-term loading.

The values (b) and (c) also depend on the sign of the bending moment. In principle, separate analyses are needed for the actions in (a), in (b), and for each relevant arrangement of variable loading in (c).

In practice, the following simplifications are made wherever possible.

(1) A value I calculated for the uncracked composite section (denoted I_1 in EN 1994-1-1) is used throughout the span. This is referred to as 'uncracked' analysis.
(2) A single value of I, based on a modular ratio that is approximately $\frac{1}{2}[(E_a/E_c') + (E_a/E_{cm})]$, is used for analyses of both types (b) and (c).
(3) Where all spans of the beam have cross-sections in Class 1 or 2 only, the influence of method of construction is neglected in analyses for ultimate limit states only, and actions applied to the steel member alone are included in analyses of type (b).

Separate analyses of type (c) are always needed for different arrangements of variable loading. It is often convenient to analyse the member for unit distributed loading on each span in turn, and then obtain the moments and shears for each load arrangement by scaling and combining the results.

The alternative to 'uncracked' analysis is to use in regions where the slab is cracked a reduced value of I (denoted I_2 in EN 1994-1-1), calculated neglecting concrete in tension but including its reinforcement. This is known as 'cracked' analysis. Its weakness is that there is no simple or accurate method for deciding which parts of each span are 'cracked'. They are different for each load arrangement, and are modified by the effects of tension stiffening, previous loadings, temperature, creep, shrinkage and longitudinal slip. A common assumption is that 15% of each span, adjacent to each internal support, is 'cracked'.

In practice, 'uncracked' analysis is usually preferred for ultimate limit states, with allowance for cracking by redistribution of moments. Deflections should be estimated with allowance for cracking, as explained in Section 4.3.2.3.

4.3.2.1 *Redistribution of moments in continuous beams*

Redistribution is a well-established and simple approximate method for modifying the results of an elastic global analysis. It allows for the inelastic behaviour that occurs in all materials in a composite beam before maximum load is reached, and for the effects of cracking of concrete at serviceability limit states. It is also used in the analysis of beams and frames of structural steel and of reinforced concrete, with limitations appropriate to the material and type of member.

It consists of modifying the bending-moment distribution found for a particular loading while maintaining equilibrium between the actions (loads) and the bending moments. Moments are reduced at cross-sections where

Table 4.3   Limits to redistribution of hogging moments, per cent of the initial
value of the bending moment to be reduced

Class of cross-section in hogging moment region	1	2	3	4
For 'uncracked' elastic analysis	40	30	20	10
For 'cracked' elastic analysis	25	15	10	0

the ratio of action effect to resistance is highest (usually, at the internal
supports). The effect is to increase the moments of opposite sign (usually,
in the mid-span regions).

For continuous composite beams, the ratio of action effect to resistance
is higher at internal supports, and lower at mid-span, than for most beams
of a single material, and the use of redistribution is essential for economy
in design. It is limited by the onset of local buckling of steel elements in
compression, as shown in Table 4.3, which is given for ultimate limit
states in EN 1994-1-1.

The differences between the two sets of figures show that 'uncracked'
analyses have been assumed to give hogging moments that are higher
than those from 'cracked' analyses by amounts that are respectively 12%,
13%, 9% and 10% for Classes 1 to 4 (e.g., 140/125 = 1.12 for Class 1).

The hogging moments referred to are the peak values at internal supports, which do not include supports of cantilevers (at which the moment
is determined by equilibrium and cannot be changed). Where the composite section is in Class 3 or 4, moments due to loads on the steel
member alone are excluded. The values in Table 4.3 are based on research
(e.g., Reference 40).

The use of Table 4.3, and the need for redistribution, is illustrated in
the following example. The Eurocode also allows limited redistribution
from mid-span regions to supports, but this is rare in practice.

4.3.2.2   *Example: redistribution of moments*

A composite beam of uniform section (apart from reinforcement) is continuous over three equal spans L. The cross-sections are in Class 1. For
the ultimate limit state, the design permanent load is g per unit length, and
the variable load is q per unit length, with $q = 2g$. The sagging moment of
resistance, M_{Rd}, is twice the hogging moment of resistance, M'_{Rd}. Find the
minimum required value for M_{Rd}:

(a)   by elastic analysis without redistribution;
(b)   by elastic analysis with redistribution to Table 4.3;
(c)   by rigid-plastic analysis.

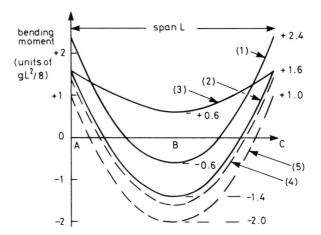

Figure 4.9 Bending moment diagrams with redistribution

For simplicity, only the middle span is considered, and only symmetrical arrangements of variable load.

Bending moment distributions for the middle span, ABC, given by 'uncracked' elastic analysis are shown in Fig. 4.9 for permanent load plus the following arrangements of variable load:

(1) q on all spans,
(2) q on the centre span only,
(3) q on the end spans only.

The moments are given as multiples of $gL^2/8$. Questions (a) to (c) are now answered.

(a) Without redistribution, the peak hogging moment, $2.4gL^2/8$, curve (1), governs the design, and since $M_{Rd} = 2M'_{Rd}$,

$$M_{Rd} \geq 4.8gL^2/8$$

(b) The peak hogging moment at each support is reduced by 40% to $1.44gL^2/8$, curve (4). The corresponding sagging moment is $(0.6 + 0.96)gL^2/8 = 1.56gL^2/8$. Elastic analysis for loadings (2) and (3) gives curves (2) and (3), respectively. Redistribution of 10% is used, so that their peak hogging moments are also $1.44gL^2/8$. This value governs the design, so that

$$M_{Rd} \geq 2.88gL^2/8$$

(c) The method used is explained in Section 4.3.3. Redistribution is unlimited, so that support moments for loading (1) are reduced by 58%, to

$1.0gL^2/8$. The corresponding sagging moment is $(0.6 + 1.4)gL^2/8$, (curve (5)). Smaller redistributions are required for the other loadings. The available resistances at the supports and at mid-span are fully used, when

$$M_{Rd} = 2.0gL^2/8$$

The preceding three results for M_{Rd} show that the resistance required is significantly reduced when the degree of redistribution is increased.

For some composite beams, use of rigid-plastic analysis can imply even larger redistribution than the 58% found here. However, design is then usually governed by serviceability criteria.

4.3.2.3 *Corrections for cracking and yielding*

Cracking of concrete and yielding of steel have less influence on deflections in service than they do on analyses for ultimate limit states, because the design loads are lower. In short cantilevers and at some internal supports there may be very little cracking, so deflections may be over-estimated by an analysis where redistribution is used as explained above. Where a low degree of shear connection is used, deflections may be increased by longitudinal slip between the slab and the steel beam.

For these reasons, design codes give modified methods of elastic analysis for the prediction of bending moments at internal supports of continuous beams. First, a method from BS 5950 for uniform beams is given that allows only for the effects of support moments. Let these hogging moments be M_1 and M_2, for a loading that would give a maximum sagging moment M_0 and a maximum deflection δ_0, if the span were simply-supported. It can be shown by elastic analysis of a uniform member with uniformly-distributed load, that the moments M_1 and M_2 reduce the mid-span deflection from δ_0 to δ_c, where

$$\delta_c = \delta_0[1 - 0.6(M_1 + M_2)/M_0] \tag{4.35}$$

This equation is quite accurate for other realistic loadings. It shows the significance of end moments. For example, if $M_1 = M_2 = 0.42M_0$, the deflection δ_0 is halved. It is not strictly correct to assume that the maximum deflection always occurs at mid-span but the error is negligible.

For cracking and yielding, the methods of EN 1994-1-1 are now described, followed by those of BS 5950. Shear lag has little effect on deflections, but section properties based on the effective flange width are often used, as they are needed for other calculations.

The general method given for allowing for the effects of cracking at internal supports is for beams with all ratios of span lengths, and is applicable for both serviceability and ultimate limit states. It involves two stages of calculation. The 'uncracked' flexural stiffness $E_a I_1$ is needed for each span, and also the 'cracked' flexural stiffness $E_a I_2$ at each internal support.

For the load arrangement considered, the bending moments due to the load applied to the composite member are first calculated using stiffnesses $E_a I_1$. At each internal support, the maximum tensile stress in the concrete due to the relevant moment is calculated. This is repeated for other relevant load arrangements. Let the highest tensile stress thus found, at a particular support, be σ_{ct}. If this stress exceeds $2f_{ctm}$ (where f_{ctm} is the mean tensile strength of the concrete), the stiffness $E_a I_1$ is replaced by $E_a I_2$ over 15% of the length of the span on each side of that support.

The analyses for bending moments are then repeated using the modified stiffnesses, and the results are used whether the new values σ_{ct} exceed $2f_{ctm}$, or not. This method is based on one that has been used for composite bridge beams since the 1960s.

EN 1994-1-1 gives an alternative to re-analysis of the structure, applicable for beams with critical sections in Class 1, 2 or 3. It is that at every support where $\sigma_{ct} > 1.5f_{ctm}$, the bending moment is multiplied by a reduction factor f_1, and corresponding increases are made in the sagging moments in adjacent spans. For general use, $f_1 = 0.6$, but a higher value, given by

$$f_1 = (E_a I_1/E_a I_2)^{-0.35} \geq 0.6 \tag{4.36}$$

is permitted for internal spans with equal loadings and approximately equal length.

Where unpropped construction is used and a high level of redistribution (e.g., 40%) is made in global analyses for ultimate limit states, it is likely that serviceability loads will cause local yielding of the steel beam at internal supports. In design to EN 1994-1-1, allowance may be made for this by multiplying the moments at relevant supports by a further factor f_2, where:

$f_2 = 0.5$, if f_y is reached before the concrete slab has hardened;
$f_2 = 0.7$, if f_y is reached due to extra loading applied after the concrete has hardened.

These methods are used in the example in Section 4.6.5.

In BS 5950, the simplified methods given are based on global analyses where the 'uncracked' stiffnesses $E_a I_1$ are used, and variable load is present on all spans. The hogging moments so found are reduced by empirical factors that take account of other arrangements of variable load.

Local yielding of the steel beam, if it occurs, causes an additional permanent deflection. This is referred to as 'shakedown' in BS 5950, and is allowed for by further reducing the hogging moments at the supports.

The calculation of deflections, with allowance for the effects of slip, is treated in Section 4.4.

4.3.3 *Rigid-plastic analysis*

For composite beams, use of rigid-plastic analysis can imply even larger redistributions of elastic moments than the 58% found in the example of Section 4.3.2.2, particularly where spans are of unequal length, or support concentrated loads.

Redistribution results from inelastic rotations of short lengths of beam in regions where 'plastic hinges' are assumed in the theory. Rotation may be limited either by crushing of concrete or buckling of steel. Rotation capacity depends on the proportions of the relevant cross-sections, as well as on the shape of the stress–strain curves for the materials.

The formation of a collapse mechanism is a sequential process. The first hinges to form must retain their resistance while undergoing sufficient rotation for plastic resistance moments to be reached at the locations where the last hinges form. Thus, the *rotation capacity* at each hinge location must exceed the *rotation required*. Neither of these quantities is easily calculated, so the requirement is in practice replaced by limitations on the use of the method, based on research. Those given in the Eurocodes include the following.

(1) At each plastic hinge location:
  - lateral restraint to the compression flange should be provided;
  - the effective cross-section should be in Class 1;
  - the cross-section of the steel component should be symmetrical about the plane of its web.
(2) All effective cross-sections in the member should be in Class 1 or Class 2.
(3) Adjacent spans should not differ in length by more than 50% of the shorter span.
(4) End spans should not exceed 115% of the length of the adjacent span.
(5) The member should not be susceptible to lateral-torsional buckling (i.e., $\bar{\lambda}_{LT} \leq 0.4$).
(6) In any span L, where more than half of the design load is concentrated within a length of $L/5$, at any sagging hinge, not more than 15% of the overall depth of the member should be in compression, unless it can be shown that the hinge will be the last to form in that span.

The method of analysis is well known, being widely used for steel-framed structures, so only an outline is given here. The principal assumptions are as follows.

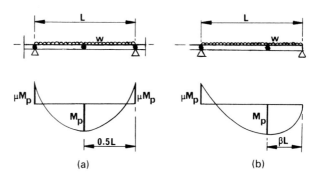

Figure 4.10 Rigid-plastic global analysis

(1) Collapse (failure) of the structure occurs by rotation of plastic hinges at constant bending moment, all other deformations being neglected.

(2) A plastic hinge forms at any cross-section where the bending moment due to the actions reaches the bending resistance of the member.

(3) All loads on a span increase in proportion until failure occurs, so the loading can be represented by a single parameter.

The value of this parameter at collapse is normally found by assuming a collapse mechanism, and equating the loss of potential energy of the loads, due to a small movement of the mechanism, with the energy dissipated in the plastic hinges.

For a beam of uniform section with distributed loading w per unit length, the only properties required are the moments of resistance at mid-span, M_p say, and at the internal support or supports, M'_p. Let

$$M'_p/M_p = \mu \tag{4.37}$$

If the beam is continuous at both ends, Fig. 4.10(a), hinges occur at the ends and at mid-span, and

$$(1 + \mu)M_p = wL^2/8 \tag{4.38}$$

If the beam is continuous at one end only, the bending moment diagram at collapse is as shown in Fig. 4.10(b). It can easily be shown that

$$\beta = (1/\mu)[\sqrt{(1 + \mu)} - 1] \tag{4.39}$$

and

$$M_p = w\beta^2L^2/2 \tag{4.40}$$

4.4   Stresses and deflections in continuous beams

Values of bending stresses at serviceability limit states may be needed in calculations for control of load-induced cracking of concrete (Section 4.2.5.3), and for prediction of deflections where unpropped construction is used. Bending moments are determined by elastic global analysis (Section 4.3.2). Those at internal supports are then modified to allow for cracking and yielding (Section 4.3.2.3). Stresses are found as in Section 3.5.3 for sagging moments, or Section 4.2.1 for hogging moments.

Deflections are much less likely to be excessive in continuous beams than in simply-supported spans, but they should always be checked where design for ultimate limit states is based on rigid-plastic global analysis. For simply-supported beams, the increase in deflection due to the use of partial shear connection can be neglected in certain circumstances (Section 3.7.1) and can be estimated from Equation 3.94. These same rules can be used for continuous beams, where they are a little conservative because partial shear connection is used only in regions of sagging bending moment.

The influence of shrinkage of concrete on deflections is treated in Section 3.8. For continuous beams, the method of calculation is rather complex, because shrinkage causes bending moments as well as sagging curvature; but its influence on deflection is much reduced by continuity.

4.5   Design strategies for continuous beams

Until experience has been gained, the design of a continuous beam may involve much trial and error. There is no ideal sequence in which decisions should be made, but the following comments on this subject may be useful.

It is assumed that the span and spacing of the beams is known, that the floor or roof slab spanning between them has been designed, and that most or all of the loading on the beams is uniformly-distributed, being either permanent (g) or variable (q). The beams add little to the total load, so g and q are known.

One would not be designing a continuous beam if simply-supported spans were satisfactory, so it can be assumed that simple spans of the maximum available depth are too weak, or deflect or vibrate too much; or that continuity is needed for seismic design, or to avoid wide cracks in the slab, or for some other reason.

The provision for services must be considered early. Will the pipes and ducts run under the beams, through the holes in the webs, or above the slab? Heavily-serviced buildings needing special solutions (castella beams, stub girders, haunched beams, etc.) are not considered here. The provision of holes in webs of continuous beams is easiest where the ratio q/g is low [32], and a low q/g is also the situation where the advantages of continuity over simple spans are greatest.

Continuity is more advantageous in beams with three or more spans than where there are only two; and end spans should ideally be shorter than interior spans. The least benefit is probably obtained where there are two equal spans. The example in Section 4.6 illustrates this. Using a steel section that could span 9.0 m simply-supported, it is quite difficult to use the same cross-section for two continuous spans of 9.5 m.

A decision with many consequences is the Class of the composite section at internal supports. Two distinct strategies are now compared.

(1) Minimal top longitudinal reinforcement is provided in the slab. If the composite beam is in Class 1, rigid-plastic global analysis can be used, unless $\bar{\lambda}_{LT} > 0.4$ (Section 4.2.4). If the beam is in Class 2, the hogging bending moment will be redistributed as much as permitted, to enable good use to be made of the available bending resistance at mid-span.

(2) The reinforcement in the slab at internal supports is heavier, with an effective area at least 1% of that of the slab. The composite section will certainly be in Class 2, perhaps Class 3. Restrictions on redistribution of moments will probably cause the design hogging moments, M'_{Ed}, to increase (cf. case (1)) faster than the increase in resistance, M'_{Rd}, provided by the reinforcement, and further increase in the latter may put the section into Class 4. So the steel section may have to be heavier than for case (1), and there will be more unused bending resistance at mid-span. However, that will allow a lower degree of shear connection to be used. With higher M'_{Ed} the bending-moment diagram for lateral-torsional buckling is more adverse. Deflections are less likely to be troublesome, but the increase in the diameter of the reinforcing bars makes crack-width control more difficult.

The method of fire protection to be used may have consequences for the structural design. For example, web encasement improves the class of a steel web that is otherwise in Class 3, but not if it is in Class 2; and it improves resistance to lateral buckling.

Finally, it has to be decided whether construction will be propped or unpropped. Propped construction allows a shallower steel beam to be used – but it will be less stiff, so the dynamic behaviour may be less satisfactory. Propped construction costs more, and crack-width control is more difficult; but design is much less likely to be governed by excessive deflection.

The design presented next is based on strategy (1) above, using a lightweight-concrete slab and an encased web. This is done to illustrate methods. It may not be the best solution.

4.6 Example: continuous composite beam

4.6.1 *Data*

So that use can be made of previous work, the design problem is identical with that of Chapter 3, except that the building (Fig. 3.1) now consists of two bays each of span 9.5 m. The transverse beams at 4 m centres are assumed to be continuous over a central longitudinal wall. They are attached to columns in the outer walls, as before, by nominally-pinned joints located 0.2 m from the centres of the columns. No account is taken of the width of the central wall. Thus, each beam is as shown in Fig. 4.11.

The use of continuity should offset the increase in span from 9 m to 9.5 m, so it is assumed initially that the designs of the slab and the mid-span region of the beam are as before, with the same materials, loads, and partial safety factors. The design loads per unit length of beam, represented by the general symbol w, and the corresponding values of the bending moment $wL^2/8$ for a span of 9.3 m are as given in Table 4.4.

Table 4.4 Loads and bending moments for a span of 9.3 m

	Characteristic loads		Ultimate loads	
	Load (kN/m)	$wL^2/8$ (kN m)	Load (kN/m)	$wL^2/8$ (kN m)
Permanent, on steel beam	12.4	134	16.7	181
Permanent, composite	5.2	56	7.0	76
Variable, composite	24.8	268	37.2	402
Total	42.4	458	60.9	659

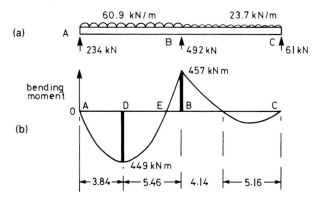

Figure 4.11 Continuous beam with dead load, plus imposed load on span AB only

Other design data from Chapter 3 are as follows:

structural steel: $f_y = 355$ N/mm^2 $f_y/\gamma_A = 355$ N/mm^2
concrete: $f_{ck} = 25$ N/mm^2 $f_{ck}/\gamma_C = 16.7$ N/mm^2
bar reinforcement: $f_{sk} = 500$ N/mm^2 $f_{sk}/\gamma_S = 435$ N/mm^2
welded fabric: $f_{sk} = 500$ N/mm^2 $f_{sk}/\gamma_S = 435$ N/mm^2
shear connectors: $P_{Rk} = 75.2$ kN $P_{Rk}/\gamma_V = 60.2$ kN

profiled steel sheeting: details as shown in Fig. 3.9; nominal thickness, 0.9 mm

composite slab: 150 mm thick, with T8 bars at 150 mm (top) above the steel beams, and at 300 mm (bottom), shown in Fig. 3.12, and concrete of Grade LC25/28

composite beam: steel section 406 × 178 UB 60, shown in Fig. 3.29, with shear connection as in Fig. 3.30 and encased web as in Fig. 3.31, using Grade C25/30 concrete. The properties of the concretes are given in Table 1.4.

Many properties of the composite cross-section are required, so it is useful to assemble them in a table, for ease of reference. The elastic and plastic properties for major-axis bending, used in Chapter 3 or in this chapter, are given in Table 4.5. The values of A and I for transformed cross-sections are based on $E = 210$ kN/mm^2. Values headed 'reinforced' are for a cross-section with 679 mm^2 of top longitudinal reinforcement. As explained later, the thickness of concrete slab above the sheeting is taken as 95 mm for elastic properties and as 80 mm for plastic properties, with the exception shown in Table 4.5. The depths x_c are from the top of the concrete slab.

Table 4.5   Properties of cross-sections of a composite beam

Type of section	Cracked unreinforced (*I*-section only)	Uncracked unreinforced				Cracked reinforced
Effective breadth, b_{eff}, mm	—	2250	2250	1225	1225	1225
Modular ratio, n	—	10.1	20.2	10.1	20.2	—
Transformed cross-section, A, mm^2	7600	28 760	18 180	17 300	12 450	8279
Depth x_c of elastic neutral axis, mm	353	129 ($h_c = 95$)	176 ($h_c = 95$)	177 ($h_c = 80$)	231 ($h_c = 80$)	327
Second moment of area, $10^{-6}I$, mm^4	215	751	636	—	508	278
Depth x_c of plastic neutral axis, mm	353	151 ($h_c = 80$)	151 ($h_c = 80$)	—	—	300
$M_{pl,Rd}$, kN m	424	829 sagging	829 sagging	—	—	510 hogging

The mid-span effective breadth of 2250 mm was found for a 8.6-m span. The increase for a 9.3-m span has little effect on the properties and has been ignored.

Other properties, not in Table 4.5, are as follows:

for the steel section:

$$10^{-6}W_{a,pl} = 1.194 \text{ mm}^3 \quad 10^{-6}I_{az} = 12.0 \text{ mm}^4 \quad V_{pl,Rd} = 697 \text{ kN}$$

4.6.2   *Flexure and vertical shear*

A rough check on the adequacy of the assumed beam section is provided by rigid-plastic global analysis. The value of $wL^2/8$ that can be resisted by each span is a little less than

$$M_{pl,Rd} + 0.5M_{pl,a,Rd} = 829 + 212 = 1041 \text{ kN m}$$

neglecting the reinforcement in the slab at the internal support, B in Fig. 4.11. This is well above $wL^2/8$ for the loading (659 kN m, from Table 4.4).

Minimum reinforcement at the internal support
It is assumed that the exposure class is X0 or XC1 (Section 4.2.5) and that the limiting crack width is 0.4 mm under quasi-permanent loading. It is assumed initially that the top longitudinal reinforcement at support B is six T12 bars ($A_s = 679 \text{ mm}^2$) at 200 mm spacing. The cross-section is then as shown in Fig. 4.1. It was found in Section 4.2.1.2 that its hogging resistance is $M_{pl,Rd} = 510 \text{ kN m}$, with effective flange width $b_{eff} = 1.225 \text{ m}$.

The area A_s may be governed by the rules for minimum reinforcement (Section 4.2.5.2). These require calculation of the distance z_0 in Fig. 4.14(a) for the uncracked unreinforced composite section with $n = 10.1$. Initial cracking is likely to occur above the small top ribs of the sheeting, where the slab thickness is 80 mm, so the assumption $h_c = 95 \text{ mm}$, made for serviceability checks at mid-span, is not appropriate.

The transformed area of the uncracked section is

$$A = 7600 + 1225 \times 80/10.1 = 17\ 300 \text{ mm}^2$$

Taking moments of area about the top of the slab for the neutral-axis depth x_c,

$$17\ 300x_c = 7600 \times 353 + 9700 \times 40 \quad \text{whence} \quad x_c = 177 \text{ mm}$$

Hence,

$$z_0 = 177 - 40 = 137 \text{ mm}$$

From Equation 4.32,

$$k_c = 1/(1 + 80/274) + 0.3 = 1.07 \quad \text{but} \leq 1.0$$

The elastic neutral axis is below the slab, so

$$A_{ct} = 1225 \times 80 = 98\,000 \text{ mm}^2$$

The stress $f_{ct,eff}$ is taken as the mean 28-day tensile strength of the Grade LC 25/28 concrete, given in Table 1.4 as 2.32 N/mm².

As a guessed area of longitudinal reinforcement is being checked, it is simplest next to use Expression 4.33 as an equality to calculate σ_s. Hence,

$$\sigma_s = 0.8k_c f_{ct,eff} A_{ct}/A_s = 0.8 \times 1 \times 2.32 \times 98\,000/679 = 268 \text{ N/mm}^2$$

The characteristic crack width is now found. Interpolating on Fig. 4.7, the maximum bar diameter for $w_k = 0.4$ mm and $\sigma_s = 268$ N/mm² is $\phi^* = 17.2$ mm. This is for the reference concrete strength $f_{ct,0} = 2.9$ N/mm². The correction for concrete strength gives

$$\phi_{max} = 17.2 \times 2.32/2.9 = 13.8 \text{ mm}$$

The 12-mm bars proposed should therefore control crack widths from imposed deformation to better than 0.4 mm.

A similar calculation finds that, for $w_k = 0.3$ mm, $\phi < 10.6$ mm, so that 10-mm bars would be needed.

Classification of cross-sections

It is easily shown by the methods of Section 4.2.1 that the steel compression flange is in Class 1 at support B. Calculations in Section 4.2.1.2 that ignored the web encasement found that the web was in Class 2, and that $M_{pl,Rd} = 510$ kN m. Web encasement enables a Class 3 web to be upgraded to Class 2; but promotion from Class 2 to Class 1 is not permitted, because crushing of the encasement in compression may reduce the rotation capacity in hogging bending to below that required of a Class 1 section. It is therefore not possible to use plastic global analysis, but $M_{pl,Rd}$ can be used as the bending resistance at support B, subject to checks for vertical shear and lateral buckling.

Vertical shear

The maximum vertical shear occurs at support B (Fig. 4.11) when both spans are fully loaded. Ignoring the effect of cracking of concrete (which reduces the shear at B) enables results for beams of uniform section to be used. From elastic theory,

$$V_{\text{Ed,B}} = 5wL/8 = 5 \times 60.9 \times 9.3/8 = 354 \text{ kN}$$

From Equation 3.113,

$$V_{\text{pl,a,Rd}} = 697 \text{ kN} \quad \text{so} \quad V_{\text{Ed}}/V_{\text{Rd}} = 0.51$$

This exceeds 0.5, so $M_{\text{pl,Rd}}$ should be reduced; but, from Equation 4.13, the reduction is obviously negligible.

Redistribution of bending moment from B reduces $V_{\text{Ed,B}}$, but shear resistance of a beam may conservatively be checked ignoring this, as the redistribution may not occur; for example, because the top reinforcement is stronger than has been assumed in design. However, in global analyses leading to column design, the design vertical shears should be consistent with the design bending moments.

Bending moments

Ignoring cracking, the maximum hogging bending moment at B occurs when both spans are fully loaded, and is

$$M_{\text{Ed,B}} = wL^2/8 = 60.9 \times 9.3^2/8 = 658 \text{ kN m}$$

From Table 4.3, the maximum permitted redistribution for a Class 2 member is 30%, giving

$$M_{\text{Ed,B}} = 658 \times 0.7 = \textbf{461 kN m}$$

which is below $M_{\text{pl,Rd}}$ (510 kN m, from Section 4.2.1.2).

The maximum sagging bending moment in span AB occurs with minimum load on span BC. Elastic analysis neglecting cracking gives the results shown in Fig. 4.11. There is no need for redistribution from support B. The maximum sagging moment, $M_{\text{Ed}} = 449$ kN m, is lower than that given by Equation 3.108, 563 kN m, for the simply-supported beam; and the vertical shear at A is lower than that at B.

4.6.3 Lateral buckling

The lateral stability of the steel bottom flange adjacent to support B is checked using the 'continuous U-frame' model explained in Section 4.2.4 and the bending-moment distribution shown for span BC in Fig. 4.11(b).

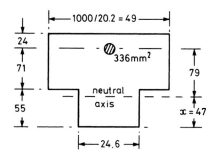

Figure 4.12 Cracked section of composite slab, for hogging bending

If it can be shown that the UB steel section used here qualifies for the relaxation given by Equation 4.28, then no check on lateral buckling is needed. It can be shown to qualify, by a new method given elsewhere [17]; but its resistance to lateral buckling is now determined, to illustrate the method of Equations 4.22 to 4.24.

In Equation 4.22,

$$M_{cr} \approx (k_c C_4/\pi)(k_s E_a I_{afz})^{1/2} \tag{4.22}$$

the term k_s represents the stiffness of the U-frame:

$$k_s = k_1 k_2/(k_1 + k_2) \tag{4.18}$$

Equation 4.21 gives k_2 for a concrete-encased web. It is assumed that the normal-density encasement has a modular ratio of 20.2 for long-term effects, so that

$$k_2 = \frac{E_a t_w b_f^2}{16 h_s(1 + 4 n t_w/b_f)} = \frac{210\,000 \times 7.8 \times 178^2}{16 \times 393(1 + 80.8 \times 7.8/178)} = 1.82 \times 10^6 \text{ N}$$

From Equation 4.19, $k_1 = 4E_d I_2/a$, where $E_a I_2$ is the 'cracked' stiffness of the composite slab in hogging bending. To calculate I_2 the trapezoidal rib shown in Fig. 3.9 is replaced by a rectangular rib of breadth $162 - 13 = 149$ mm. Using a modular ratio $n = 20.2$, the transformed width of rib is $149/(0.3 \times 20.2) = 24.6$ mm per metre width of slab, since the ribs are at 0.3 m spacing. The transformed section is thus as shown in Fig. 4.12. Reinforcement within the rib (Fig. 3.12) is neglected.

The position of the neutral axis is given by

$$\tfrac{1}{2} \times 24.6x^2 = 336(126 - x) \quad \text{whence} \quad x = 47 \text{ mm}$$

then

$$10^{-6} I_2 = 336 \times 0.079^2 + 24.6 \times 0.47^3/3 = 2.95 \text{ mm}^4/\text{m}$$

From Equation 4.19, with the beam spacing $a = 4.0$ m,

$$k_1 = 4 \times 210\,000 \times 2.95/4 = 0.619 \times 10^6 \text{ N}$$

From Equation 4.18,

$$k_s = 0.619 \times 1.82 \times 10^6/2.44 = 0.463 \times 10^6 \text{ N}$$

For the steel bottom flange,

$$I_{afz} = b_f^3 t_f/12 = 178^3 \times 12.8/12 = 6.01 \times 10^6 \text{ mm}^4$$

Factor k_c, Equation 4.16, is concerned with stiffness, so the appropriate depth of slab, h_c, is 95 mm, not 80 mm. The following values are now required (see Fig. 4.5):

$$h_s = 406 - 13 = 393 \text{ mm} \quad z_c = 353 - 95/2 = 306 \text{ mm}$$

also,

$$A = A_a + A_s = 7600 + 679 = 8279 \text{ mm}^2$$

From Equation 4.17,

$$e = \frac{AI_{ay}}{A_a z_c (A - A_a)} = \frac{8279 \times 215.1 \times 10^6}{7600 \times 306 \times 679} = 1128 \text{ mm}$$

The second moment of area of the cracked composite section is found in the usual way, Section 4.2.1.3. The elastic neutral axis is 26 mm above the centroid of the steel section, and

$$10^{-6} I_y = 215 + 7600 \times 0.026^2 + 679 \times 0.292^2 = 278 \text{ mm}^4$$

From Equation 4.16,

$$k_c = \frac{h_s I_y/I_{ay}}{\dfrac{h_s^2/4 + (I_{ay} + I_{az})/A_a}{e} + h_s}$$

$$= \frac{393 \times 278/215.1}{\left[\dfrac{393^2/4 + (215 + 12) \times 10^6/7600}{1128}\right] + 393} = 1.12$$

Where A_s/A_a is small, it is simpler, and conservative, to take k_c as 1.0.

Factor C_4 is now found, using Fig. 4.6. From Fig. 4.11, for span BC,

$$M_0 = 23.7 \times 9.3^2/8 = 256 \text{ kN m} \quad \text{so} \quad \psi = 457/256 = 1.78$$

From Fig. 4.6, $C_4 = 17.4$.

The effect of including in Equation 4.15 the St Venant torsion constant for the steel section is negligible, so the simpler Equation 4.22 can be used. It gives:

$$M_{cr} = (1.12 \times 17.4/\pi)(0.463 \times 210\ 000 \times 6.01)^{1/2} = \textbf{4742 kN m}$$

The characteristic plastic bending resistance at support B is required for use in Equation 4.23. The design value was found in Section 4.2.1.2 to be 510 kN m. Replacing the γ_S factor for reinforcement (1.15) by 1.0 increases it to $M_{pl,Rk} = 579$ kN m.

The slenderness $\bar{\lambda}_{LT}$ is given by Equation 4.23, which is

$$\bar{\lambda}_{LT} = (M_{pl,Rk}/M_{cr})^{1/2} = (579/4742)^{1/2} = 0.35$$

This is less than 0.4, so $M_{pl,Rd}$ need not be reduced to allow for lateral buckling.

4.6.4 *Shear connection and transverse reinforcement*

For sagging bending, the resistance required, 449 kN m, is well below the resistance with full shear connection, 829 kN m, so the minimum degree of shear connection may be sufficient. For spans of 9.3 m, and an effective span of $9.3 \times 0.85 = 7.9$ m, this is given by Fig. 3.19 as

$$n/n_f \geq 0.49$$

In Equation 3.67, $N_c/N_{c,f}$ may be replaced by n/n_f. This equation for the interpolation method (Fig. 3.16) then gives

$$M_{Rd} = M_{pl,a,Rd} + (n/n_f)(M_{pl,Rd} - M_{pl,a,Rd}) = 424 + 0.49(847 - 424)$$

$$= \textbf{631 kN m}$$

which is sufficient.

From Equation 3.116, the resistances of the stud connectors are 51.0 kN and 42.0 kN, for one and two studs per rib, ($n_r = 1$ and $n_r = 2$, respectively). From Section 3.11.1, $N_{c,f} = 2555$ kN, so

$$N_c = 0.49 \times 2555 = 1250 \text{ kN}$$

There are 13 troughs at 0.3 m spacing in the 3.84-m length AD in Fig. 4.11. The choice of n_r is discussed in Section 3.11.2. Assuming that $n_r \geq 2$, the number of studs needed is 1250/42 = 30, so four studs are provided in each of the two troughs nearest to support A, and two in each of the other 11 troughs; total, 30.

From Fig. 4.11, the length DB is 5.46 m. Its sheeting has 5.46/0.3 = 18 troughs. The force to be resisted is 1250 kN plus 295 kN (Section 4.2.1.2) for the reinforcement at cross-section B, at yield; total, 1545 kN, requiring 37 studs at 42 kN each. The provision of four studs in the trough nearest to support B, and two in each of the other 17 gives 38 studs, which is sufficient.

The transverse reinforcement should be as determined for the sagging region in Chapter 3.

4.6.5 *Check on deflections*

The limits to deflections discussed in Section 3.7.2 correspond to the characteristic combination of loading (Expression 1.8). Where there is only one type of variable load, as here, this is simply $g_k + q_k$, but three sets of calculations may be required, because part of the permanent load g acts on the steel member and part on the composite member, and two modular ratios are needed for the composite member.

In practice, it is usually accurate enough to combine the two calculations for the composite member, using a mean value of the modular ratio (e.g., $n = 20.2$ here).

For design purposes, maximum deflection occurs when the variable load is present on the whole of one span, but not on the other span. The three loadings are shown in Fig. 4.13, with the bending-moment distributions given by 'uncracked' elastic analyses, in which the beam is assumed

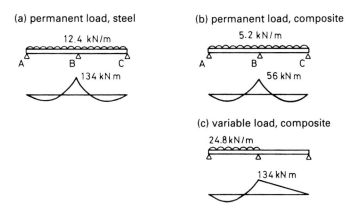

Figure 4.13 Loading for deflection of span AB

Table 4.6 Calculation of maximum deflection

	w (kN/m)	M_0 (kN m)	M_B (kN m)	f_1	f_2	M_1 (kN m)	$10^{-6}I_1$ (mm^4)	δ_c (mm)
g on steel	12.4	134	134	—	—	134	215	10.7
g on composite	5.2	56	56	0.6	0.7	24	636	2.8
q on composite	24.8	268	134	0.6	0.7	56	636	15.9

to be of uniform section. The data and results are summarised in Table 4.6, where M_B is the hogging moment at support B at this stage of the analysis.

Following the method of Section 4.3.2.3, the maximum tensile stress in the uncracked composite section at B, σ_{ct}, is now found, using an effective width of 1.225 m, Fig. 4.14(a). This stress will occur when variable load acts on both spans, and is calculated using $n = 20.2$ for all load on the composite member. Using data from Table 4.5,

$$\sigma_{ct} = \Sigma\left(\frac{Mx}{nI_1}\right) = \frac{(56 + 268) \times 231}{20.2 \times 508} = 7.3 \text{ N/mm}^2$$

where nI_1 is the 'uncracked' second moment of area in 'concrete' units.

This stress exceeds $1.5f_{lctm}$ (3.48 N/mm^2). To avoid re-analysis with spans of non-uniform section, the correction factor f_1 given by Equation 4.36 can be used. For these 'external' spans, $f_1 = 0.6$.

The maximum compressive stress in the steel bottom fibre is now calculated, to determine whether the correction factor f_2 for yielding is

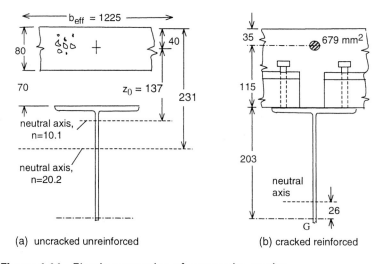

(a) uncracked unreinforced   (b) cracked reinforced

Figure 4.14 Elastic properties of composite section

required. As for σ_{ct}, variable load should be assumed to act on both spans. Using data from Section 4.6.1 and Fig. 4.14(b),

$$\sigma_{4,a} = \sum \left(\frac{Mx_4}{I_2} \right) = \frac{134 \times 203}{215} + \frac{(56 + 268) \times 229}{278}$$

$$= 126 + 267 = 393 \text{ N/mm}^2$$

where x_4 is the distance of the relevant neutral axis above the bottom fibre. The result shows that yielding occurs (393 > 355), but not until after the slab has hardened (126 < 355), so from Section 4.3.2.3, $f_2 = 0.7$. The hogging moments M_1 for use in Equation 4.35 are

$$M_1 = f_1 f_2 M_B$$

and are given in Table 4.6. The other end moment, M_2, is zero.

Deflections δ_0 for each loading acting on a simply-supported span are now required. These are, in general:

$$\delta_0 = \frac{5wL^4}{384EI} = \frac{5 \times 9.3^4 \times 10^9}{384 \times 210} \left(\frac{w}{I_1} \right) = 464 \times 10^6 (w/I_1) \text{ mm}$$

with w in kN/m and I_1 in mm^4. Using values from Table 4.6 in Equation 4.35 gives the total deflection:

$$\delta_c = 464 \times 10^6 \, \Sigma[(w/I_1)(1 - 0.6M_1/M_0)]$$

$$= 464[(12.4/215) \times 0.4 + (5.2/636)(1 - 0.6 \times 24/56)]$$

$$+ 464[(24.8/636)(1 - 0.6 \times 56/268)]$$

$$= 10.7 + 2.8 + 15.9 = 29.4 \text{ mm}$$

This result is probably too high, because the factor f_1 may be conservative, and no account has been taken of the stiffness of the web encasement. This total deflection is span/316, less than the guideline of $L/300$ given in the UK's national annex to EN 1990 [12], for floors with plastered ceilings and/or non-brittle partitions.

The total deflection of the simply-supported span of 8.6 m, for the same loading, was found to be 35.5 mm. This would be $35.5 \times (9.3/8.6)^4 =$ 48.5 mm for the present span. The use of continuity at one end has reduced this value by 19.1 mm, or 39%.

Even so, these deflections are fairly large for a continuous beam with a ratio of span to overall depth of only $9300/556 = 16.7$. This results from the use of unpropped construction, high-yield steel and lightweight concrete.

Deflections of a similar propped structure in mild steel and normal-density concrete would be much lower.

4.6.6 *Control of cracking*

The widest cracks will occur at the top surface of the slab, above an internal support. The reinforcement at this cross-section is 12-mm bars at 200 mm spacing ($A_s = 679$ mm^2). It was shown in Section 4.6.2 that this can control the widths of cracks from imposed deformation to below 0.4 mm. The crack width caused by the characteristic loading is now found, using Section 4.2.5.3.

The bending moments M_B given in Table 4.6 are applicable, except that M_B for imposed load must be doubled, as for this purpose it acts on both spans. For deflections, use of the reduction factor $f_1 = 0.6$ probably under-estimated M_B. For checking crack width, any approximation should be an over-estimate. In Table 4.3 for limits to redistribution of moments, it is assumed that cracking causes a 15% reduction in a Class 2 section, so f_1 is taken as 0.85.

The factor f_2 for yielding of steel is not applied, as the yield strength of the steel is likely to be higher than specified. Hence, for cracking,

$$M_B = (56 + 2 \times 134) \times 0.85 = 275 \text{ kN m}$$

For the cracked composite section, $10^{-6}I_2 = 278$ mm^4, from Table 4.5, so the stress in the reinforcement, at distance 292 mm above the neutral axis, is

$$\sigma_{s,o} = 275 \times 292/278 = 289 \text{ N/mm}^2$$

This must be increased to allow for tension stiffening between cracks. From Section 4.2.5.3,

$$\alpha_{st} = AI_2/A_aI_a = 8279 \times 278/(7600 \times 215) = 1.41$$

From Section 4.6.2, the cracked area of concrete is

$$A_{ct} = b_{eff}h_c = 1225 \times 80 = 98\,000 \text{ mm}^2$$

From Equation 4.34, with f_{lctm} from Table 1.4,

$$\sigma_s = \sigma_{s,o} + 0.4f_{lctm}A_{ct}/(\alpha_{st}A_s)$$
$$= 289 + 0.4 \times 2.32 \times 98\,000/(1.41 \times 679) = 384 \text{ N/mm}^2$$

This is below the yield strength, so the existing 12-mm bars are satisfactory if a non-brittle floor finish is to be used, such that cracks will not be visible.

If, however, control of crack widths to 0.4 mm were required, reference to Fig. 4.7 shows that 8-mm bars would be required. To provide 679 mm^2, their spacing would have to be 90 mm, or bars in pairs at 180 mm.

The alternative of increasing the top reinforcement to, say, 10-mm bars at 100 mm (883 mm^2) would require some re-calculation. It increases the ultimate hogging bending moment $M_{Ed,B}$ and so makes susceptibility to lateral buckling more likely. It is not obvious whether it would reduce the value of the tensile stress σ_s to below 360 N/mm^2, the limit given in both Figs 4.7 and 4.8 for crack control to 0.4 mm. This illustrates interactions that occur in design of a hogging moment region.

4.7 Continuous composite slabs

The concrete of a composite slab floor is almost always continuous over the supporting beams, but the individual spans are often designed as simply-supported (Sections 3.3 and 3.4), for simplicity. Where deflections are found to be excessive, continuous design may be used, as follows.

Elastic theory is used for the global analysis of continuous sheets acting as shuttering. Variations of stiffness due to local buckling of compressed parts can be neglected. Resistance moments of cross-sections are based on tests (Section 3.3).

Completed composite slabs are generally analysed for ultimate bending moments in the same way as continuous beams with Class 2 sections. 'Uncracked' elastic analysis is used, with up to 30% redistribution of hogging moments, assuming that the whole load acts on the composite member. Rigid-plastic global analysis is allowed by EN 1994-1-1 where all cross-sections at plastic hinges have been shown to have sufficient rotation capacity. This has been established for spans less than 3.0 m with reinforcement of 'high ductility' as defined in EN 1992-1-1 for reinforced concrete. No check on rotation capacity is then required.

At internal supports where the sheeting is continuous, resistance to hogging bending is calculated by rectangular-stress-block theory, as for composite beams, except that local buckling is allowed for by using an effective width for each flat panel of sheeting in compression. This width is given in EN 1994-1-1 as twice the value specified for a Class 1 steel web, thus allowing for the partial restraint from the concrete on one side of the sheet. Where the sheeting is not continuous at a support, the section is treated as reinforced concrete.

For the control of cracking at internal supports, EN 1994-1-1 refers to EN 1992-1-1. In practice, the reinforcement to be provided may be governed by design for resistance to fire, as in Section 3.3.7, or by the transverse reinforcement required for the composite beam that supports the slab.

Chapter 5
Composite columns and frames

5.1 Introduction

A definition of *composite frame* is given in Section 4.1, with discussion of *joints*, *connections*, and of the relationship between these and the methods of global analysis (Table 4.1).

To illustrate the presentation of continuous beams in the context of buildings, it was necessary to use, in Section 4.6, an atypical example: a two-span beam with a wall as its internal support, with no transfer of bending moment between the wall and the beam.

Where a beam is supported by a column through a joint that is not 'nominally pinned', the bending moments depend on the properties of the joint, the beam and the column. They form part of a frame, which may have to provide resistance to horizontal loads such as wind (an 'unbraced frame'); or these loads may be transferred to a bracing structure by the floor slabs.

Where the lateral stiffness of the bracing structure is sufficient, the frame can be designed to resist only the vertical loads (a 'braced frame'). The lift and staircase regions of multi-storey buildings often have concrete walls, for resistance to fire. These can provide stiff lateral restraint, as can the end walls of long narrow buildings. The frames can then be designed as 'braced'.

Unbraced frames require stronger members and joints. The long-standing empirical 'wind-moment' design method [43], in current use for frames of moderate height, is not given in EN 1994-1-1.

Some of the design rules of Eurocodes 3 and 4 for semi-rigid and partial-strength joints are so recent that there is little experience of their use in practice. The scope of this chapter is limited to braced frames with beam-to-column joints that are either 'nominally pinned' or 'rigid and full strength'. The structure shown in Fig. 5.1 is used as an example. Typical plane frames such as DEF are at 4.0-m spacing, and support composite floor slabs as designed in Section 3.4. Each frame has ten two-span beams

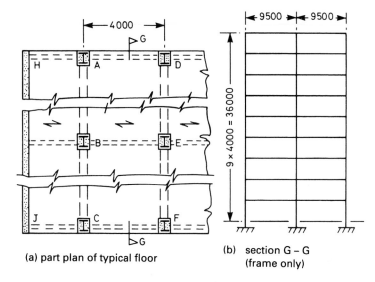

(a) part plan of typical floor

(b)   section G – G
(frame only)

Figure 5.1   A composite frame (simplified)

identical with those designed in Section 4.6, except that the wall that
provided the internal support is replaced by composite columns (B, E, etc.)
at 4.0-m spacing, with rigid and full strength beam-to-column joints.

The joints to the external columns, near points A, C, D, F, etc., are
nominally pinned. The design bending moments for the columns depend on
the assumed location of these pins, which is discussed in Section 5.4.4.2.

The building is assumed to be 60 m long. Lateral support is provided
by shear walls at each end (Fig. 5.1(a)), and a lift and staircase tower (the
'core') at mid-length, 4 m wide. Horizontal loads normal to the length of
the building are assumed to be transferred to these bracing elements by
each floor slab, spanning $(60 - 4)/2 = 28$ m between the core and an end
wall. These slabs are quite thin, but as horizontal beams they are about
19 m deep. Their span/depth ratio is so low $(28/19 = 1.47)$ that for lateral
load they are very stiff, and stresses are low.

The effect of horizontal forces in other directions is assumed to be
negligible. External walls are assumed to be supported by edge beams
such as JC and CF, spanning between the external columns. These beams
will not be designed.

This structure is used here as the basis for explanation of behaviour and
design methods, and is not fully realistic. For example, it lacks means of
escape near its two ends.

The layouts of most multi-storey buildings are such that their frames
can be designed as two-dimensional. The columns are usually designed
with their webs co-planar with those of the main beams, as shown in

Fig. 5.1(a), so that beam–column interaction causes major-axis bending in both members.

For global analysis for gravity loads, each plane frame is assumed to be independent of the others. For each storey-height column length, an axial load N_y and end moments $M_{1,y}$ and $M_{2,y}$ are found for the major-axis frame, and corresponding values N_z, $M_{1,z}$ and $M_{2,z}$ for the minor-axis frame such as HAD in Fig. 5.1. The column length is then designed (or an assumed design is checked) for axial load $N_y + N_z$ and for the bi-axial bending caused by the four end moments.

In the design of a multi-storey composite plane frame, allowance must be made for imperfections. Global imperfections, such as out-of-plumb columns, influence lateral buckling of the frame as a whole ('frame instability'). Member imperfections, such as bow of a column length between floor levels, influence the buckling of these lengths ('member instability'), and may even affect the stability of a frame.

Global analysis is usually linear-elastic, with allowance for creep, cracking and the moment–rotation properties of the joints. First-order analysis is used wherever possible, but checks must first be made that second-order effects (additional action effects arising from displacement of nodes or bowing of members) can be neglected. If not, second-order analysis is used.

A set of flow charts for the design of such a frame, given elsewhere [17], is too extensive to reproduce here. However, the sequence of these charts will be followed for the frame shown in Fig. 5.1(b), which will be found to be free from many of the complications referred to above.

Columns and joints are discussed separately in Sections 5.2 and 5.3. The Eurocode methods for analysis of braced frames are explained in Section 5.4, with a worked example. Details of the design method of EN 1994-1-1 for columns are then given, followed by calculations for two of the columns in the frame.

5.2 Composite columns

Steel columns in multi-storey buildings need protection from fire. This is often provided by encasement in concrete. Until the 1950s, it was normal practice to use a wet mix of low strength, and to neglect the contribution of the concrete to the strength and stability of the column. Tests by Faber [44] and others then showed that savings could be made by using better-quality concrete and designing the column as a composite member. This led to the 'cased strut' method of design. This was originally (in BS 449) a permissible-stress method for the steel member, which had to be of H- or I-section. It then became available in limit-state form [19]. In this method, the presence of the concrete is allowed for in two ways. It is assumed to resist a small axial load; and to reduce the effective slenderness of the steel member, which increases its resistance to axial load. Resistance to bending moment is assumed to be provided entirely by the steel. No account is taken of the resistance of the longitudinal reinforcement in the concrete.

Tests on cased struts under axial and eccentric load show that this cased strut method gives a very uneven and usually excessive margin of safety. For example, Jones & Rizk [45] quote load factors ranging from 4.7 to 6.7, and work by Faber [44] supports this conclusion. The method was improved in BS 5950, but is still generally very conservative. Its main advantage is that it is simpler than the more rational and economical methods now available.

One of the earliest methods to take proper account of the interaction between steel and concrete in a concrete-encased H-section column is due to Basu & Sommerville [46]. It has been extended to include bi-axial bending, and agrees quite well with the results of tests and numerical simulations [47, 48]. It was thought to be too complex for routine use for columns in buildings, but is included in the British code for composite bridges. Its scope includes concrete-filled steel tubes [49], which have been used as bridge piers, for example in multi-level motorway interchanges.

The Basu and Sommerville method is based on the use of algebraic approximations to curves obtained by numerical analyses. For Eurocode 4:Part 1.1, preference was given to a method developed by Roik, Bergmann and others at the University of Bochum. It has wider scope, is based on a clearer conceptual model, and is slightly simpler. It is described in Section 5.6, with a worked example.

5.3 Beam-to-column joints

5.3.1 *Properties of joints*

Three types of joint between a steel beam and the flange of an H-section steel column are shown in Fig. 5.2, and a short end-plate joint is shown in Fig. 5.22. They are all bolted, because they are made on site, where welding is expensive and difficult to inspect. The column shown in Fig. 5.2(a) is in an external wall. At an internal column, another beam would be connected to the other flange. There may also be minor-axis beams, connected to the column web as shown in Fig. 5.2(c).

Where the beams are composite and the column is internal, longitudinal reinforcement in the slab will be continuous past the column, as shown in Fig. 5.2(c). It may be provided only for the control of cracking; but if it

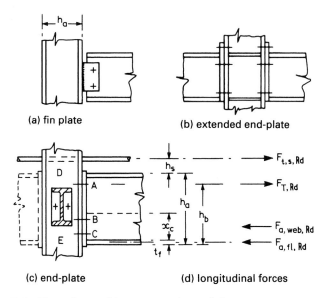

(a) fin plate

(b) extended end-plate

(c) end-plate

(d) longitudinal forces

Figure 5.2 Elevations of beam-to-column joints

consists of individual bars, rather than welded fabric, the tension in the bars may be assumed to contribute to the bending resistance of the joint, as shown in Fig. 5.2(d). Small-diameter bars may fracture before the rotation of the hogging region of the beam becomes large enough for the resistance of the joint to reach its design value, so these bars should be at least 16 mm in diameter [8].

In the fin-plate joint of Fig. 5.2(a), the bolts are designed mainly for vertical shear, and the flexural stiffness is low. The end-plate joint of Fig. 5.2(c) is likely to be 'semi-rigid' (defined later). The bolts at A are usually designed for tension only, and bolts in the compression zone (B and C) are designed for vertical shear only.

To achieve a 'rigid' connection it may be necessary to use an extended end plate and to stiffen the column web in regions D and E, as shown in Fig. 5.2(b).

Resistance of an end-plate joint
Until tabulated data become available, design of a semi-rigid partial-strength end-plate joint, as in Fig. 5.2(c), requires extensive calculations. These are explained, with an example, in the *Designers' Guide to EN 1994-1-1* [17]. An outline of the method is now given, assuming beams of the same depth on opposite sides of the column.

The rotation capacity of the joint is ensured by using a thin end plate, so that yield lines form in it before the bolts at A fracture in tension. Plastic bending of the column flange and yielding of the column web at D may also occur. The check on bolt fracture may need to allow for prying action (increase of bolt tension caused by compressive force where the edges of the end plate bear against the column flange). The tensile resistance of the top bolts $F_{T,Rd}$ is given by the weakest of these types of deformation.

The longitudinal reinforcement in the slab is assumed to be at yield in tension, so the force $F_{t,s,Rd}$ is known.

Assuming that any axial force in the beam is negligible, the compressive force at the bottom of the joint cannot exceed $F_{T,Rd} + F_{t,s,Rd}$. Failure could occur by buckling of the column web at E, so this resistance is found next, allowing for the axial compression in the column. If buckling governs, a stiffener can be added, but this is rarely necessary. The compressive force to cause yielding of the bottom flange, $F_{a,fl,Rd}$, is then found. If it is less than the total tensile force, an area of web is assumed also to yield, such that

$$F_{a,fl,Rd} + F_{a,web,Rd} = F_{T,Rd} + F_{t,s,Rd}$$

The lines of action of these four forces are known, so the bending resistance of the joint, $M_{j,Rd}$, is found.

For beams of unequal depth, or at an external column, checks are also needed on the resistance of the column web to shear and the transfer to the column of the unbalanced tensile force in the slab reinforcement. The methods can also allow for concrete encasement to the column and/or the webs of the beams.

The resistance of the joint to vertical shear is normally provided by the bolts at B and C, and may be limited by the bearing strength of the end plate or column flange. The allocation of some of the shear to the bolts at A would reduce their design resistance to tension.

Moment–rotation curve for an end-plate joint

The other information needed for design is a curve of hogging bending moment against rotation of the joint, ϕ. This is defined as the rotation additional to that which would occur if the joint were rigid and the beam continued to its intersection with the centre-line of the column, as shown in Fig. 5.3. For steel connections, methods are given in EN 1993-1-1 [15] for the prediction of this curve. They are applicable also to composite joints, with modifications given in Annex A of EN 1994-1-1. These allow for the effect of slip of the shear connection on the longitudinal stiffness of the top reinforcement, and are explained in Reference 17. The elastic properties of the components give the initial elastic stiffness, $S_{j,ini}$. This is assumed to be applicable for bending moments $M_{j,Ed}$ up to $2M_{j,Rd}/3$, where $M_{j,Rd}$ is the bending resistance of the joint.

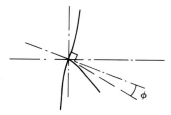

Figure 5.3 Rotation of a joint

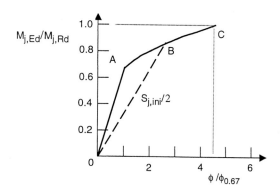

Figure 5.4 Moment–rotation curve for an end-plate joint

At higher bending moments, EN 1993-1-8 gives the stiffness as

$$S_{j,ini}/S_j = (1.5M_{j,Ed}/M_{j,Rd})^{\psi}$$

where ψ depends on the type of joint, and is 2.7 for a welded or bolted end-plate joint. The moment–rotation curve for $\psi = 2.7$ is shown as 0ABC in Fig. 5.4, in which $\phi_{0.67}$ is the rotation for $M_{j,Ed} = 2M_{j,Rd}/3$.

It is inconvenient for analysis to have a joint stiffness that depends on the bending moment. For beam-to-column joints, EN 1993-1-8 permits the simplification that for all values of $M_{j,Ed}$, $S_j = S_{j,ini}/2$. This is line OB in Fig. 5.4.

5.3.2 *Classification of joints*

As shown in Table 4.1, beam-to-column joints are classified in Eurocode 4, as in Eurocode 3, by rotational stiffness, which is relevant to elastic global analysis, and by resistance to bending moment, which is relevant to the resistance of a frame to ultimate loads. The three stiffness classes are shown in Fig. 5.5.

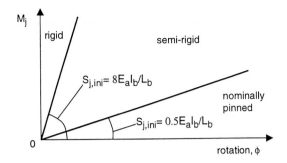

Figure 5.5 Classification of joints by initial stiffness

A *nominally pinned* joint has

$$S_{\mathrm{j,ini}} \leq 0.5 E_a I_b / L_b \qquad (5.1)$$

where $E_a I_b$ is the rotational stiffness of the connected beam, of length L_b. The value of $E_a I_b$ should be consistent with that taken for a cross-section adjacent to the joint in global analysis of the frame. The significance of this limit to $S_{\mathrm{j,ini}}$ can be illustrated by considering a beam of span L_b and uniform section that is connected at each end to rigid columns, by connections with $S_{\mathrm{j,ini}} = 0.5 E_a I_b / L_b$. It can be shown by elastic analysis that for a uniformly-distributed load w per unit length, the restraining (hogging) moments at each end of the beam are

$$M_{\mathrm{el}} = (w L_b^2 / 8) / 7.5 \qquad (5.2)$$

These end moments act also on the columns, the flexibility of which would in practice reduce the moments below M_{el}. In design with the pins on the column centre-line, it is being assumed that columns designed for $M_{\mathrm{el}} = 0$ are not 'adversely affected' by bending moment from the joint.

A joint is *rigid* if

$$S_{\mathrm{j,ini}} \geq k_b E_a I_b / L_b$$

where $k_b = 8$ for a braced frame (defined in EN 1993-1-8).

The amount of redistribution of elastic moments caused by the flexibility of a connection that is just 'rigid' can be quite significant. As an example, we consider the same beam as before, with properties $E_a I_b$ and L_b, supported at each end by rigid columns, with uniform load such that both end moments are $2M_{\mathrm{j,Rd}}/3$, when the joints have stiffness $S_{\mathrm{j,ini}} = 8 E_a I_b / L_b$.

Elastic analysis for a uniform beam shows that the end moments are $wL^2/15$. They would be $wL^2/12$ if the joints were truly rigid, so their flexibility causes a 20% redistribution of hogging moment. The situation for a composite beam in practice is more complex because $E_a I_b$ is not uniform along the span, and the columns are not rigid.

A *semi-rigid* joint has an initial rotational stiffness between these two limits, Fig. 5.5.

The classification of joints by strength is as follows.

A joint with design resistance $M_{j,Rd}$ is classified as *nominally pinned* if $M_{j,Rd}$ is less than 25% of the bending resistance of the weaker of the members joined, and if it has sufficient rotation capacity. It is not difficult to design connections that satisfy these conditions. An example is given in Section 5.10.

A *full-strength* joint has a design resistance (to bending, taking account of co-existing shear) at least equal to $M_{pl,Rd}$ for the members joined. There is a separate requirement to check that the rotation capacity of the connection is sufficient. This can be difficult. It is waived if

$$M_{j,Rd} \geq 1.2 M_{pl,Rd} \tag{5.3}$$

so in practice a 'full-strength' connection may be designed to satisfy Condition 5.3. It can then be assumed that inelastic rotation occurs in the beam adjacent to the connection. The rotation capacity is then assured by the classification system for steel elements in compression.

A *partial-strength* joint has a resistance less than that of the members joined; but must have sufficient rotation capacity, if it is at the location of a plastic hinge, to enable all the necessary plastic hinges to develop under the design loads.

5.4 Design of non-sway composite frames

5.4.1 *Imperfections*

The scope of this Section is limited to multi-storey structures of the type shown in Fig. 5.1, modelled as two sets of plane frames as explained in Section 5.1. It is assumed that the layout of the beams and columns and the design ultimate gravity loads on the beams are known.

The first step is to define the imperfections of the frame. These arise mainly from lack of verticality of columns, but also have to take account of lack of fit between members, effects of residual stresses in steelwork, and other minor influences, such as non-uniform temperature of the structure. The term 'column' is here used to mean a member that may extend over the whole height of the building. A part of it with a length equal to a storey height is referred to as a 'column length', where this is necessary to avoid ambiguity.

Imperfections within a storey-height column of length L are represented by an initial bow, e_0. This ranges from $L/100$ to $L/300$, depending on the type of column and the axis of bending. For major-axis bending of the columns shown in Fig. 5.1(a), EN 1994-1-1 specifies a bow of $L/200$, or 20 mm in 4.0 m. This has to be allowed for in verification of the member, but not in first-order global analysis. The condition of EN 1994-1-1 for neglecting member imperfections in second-order analysis is, essentially, that $N_{Ed} \leq 0.25N_{cr}$, where N_{cr} is the elastic critical axial normal force for the member, allowing for creep, given by Equation 5.20.

A bow of 20 mm in 4 m exceeds the tolerance that would be acceptable in construction because it allows also for other effects, such as residual stresses in steel. The out-of-straightness e_0 is assumed to occur at mid-length. No assumption is made about the shape of the imperfection.

Imperfections in beams are allowed for in the classification system for steel elements in compression, and in design for lateral buckling.

Frame imperfections are represented by an initial side-sway, ϕ, as shown in Fig. 5.6(a) for a single column length of height h, subjected to an axial load N. The action effects in the column are the same as if it were vertical and subjected to horizontal forces $N\phi$, as shown.

It is assumed that the angle ϕ for a composite frame is the same as for the corresponding steel frame. This is given in EN 1993-1-1 as a function

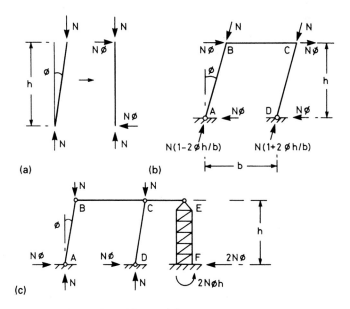

Figure 5.6 Unbraced and braced frames

of the height of the structure in metres, h, and the number of columns in the plane frame considered, m, as follows:

$$\phi = \alpha_h \alpha_m / 200 \tag{5.4}$$

where

$$\alpha_h = 2/\sqrt{h} \quad \text{with} \quad 2/3 \le \alpha_h \le 1 \tag{5.5}$$

and

$$\alpha_m = (0.5 + 1/m)^{1/2} \tag{5.6}$$

Thus, in the frame in Fig. 5.1(b), $h = 36$ m, $\alpha_h = 2/3$, $m = 3$, $\alpha_m = 0.816$, and

$$\phi = 0.67 \times 0.816/200 = 1/366$$

This initial sway applies in all horizontal directions, and is uniform over the height of the frame. In this example, the overall out-of-plumb of each column is assumed to be $36\,000/366 = 98$ mm.

Let the total design ultimate gravity load on the frame, for a particular combination of actions, be $G + Q$ per storey. The imperfections can then be represented by a notional horizontal force $\phi(G + Q)$ at each floor level – but there may or may not be an equal and opposite reaction at foundation level.

To illustrate this, we consider the single-bay single-storey unbraced frame ABCD shown in Fig. 5.6(b), with pin joints at A and D, and assume $\sin \phi = \phi$, $\cos \phi = 1$. The use of additional forces $N\phi$ at B and C is associated with the assumption that the loads N still act along the columns, as shown. There are obviously horizontal reactions $N\phi$ at A and D; but the vertical reactions N are replaced by reactions $N(1 \pm 2\phi h/b)$ at angle ϕ to the vertical. The total horizontal reaction at A is therefore

$$N\phi - N\phi(1 - 2\phi h/b) = 2N\phi^2 h/b \approx 0$$

The maximum first-order bending moment in the perfect frame is zero. The imperfection ϕ increases it to $N\phi h$, at corners B and C, which may not be negligible.

If there are pin joints at these corners, the frame has to be braced against side-sway by connection to the top of a stiff vertical cantilever EF (Fig. 5.6(c)). The external reactions now do include horizontal forces $N\phi$ at A and D, with an opposite reaction $2N\phi$ at F; and the vertical reactions at A and D are independent of ϕ.

These simple analyses are *first order*. That is, they neglect any increase in the assumed sway ϕ caused by the deformations of the structure under load. Analyses that take account of this effect are referred to as *second order*. A simple example is the elastic theory for the lateral deformation of an initially crooked pin-ended strut.

5.4.2   *Elastic stiffnesses of members*

The determination of these properties requires consideration of the behaviour of joints and of creep and cracking of concrete. Creep in beams will be allowed for by using a modular ratio $n = 2n_0 = 20.2$, as before. For columns, EN 1994-1-1 gives the effective modulus for concrete as

$$E_c = E_{cm}/[1 + (N_{G,Ed}/N_{Ed})\varphi_t] \tag{5.7}$$

where N_{Ed} is the design axial force, $N_{G,Ed}$ is the part of N_{Ed} that is permanent, and φ_t is the creep coefficient. Expressions for the stiffnesses of composite columns are given in Section 5.6.3 in terms of E_c. Typical values for the short-term elastic modulus E_{cm} are given in Section 3.2.

For cracking it will be assumed, from Section 4.3.2, that the 'cracked reinforced' section of each beam is used for a length of 15% of the span on each side of the central column. The joints are assumed to be rigid at the central column and nominally-pinned at the external columns, as explained earlier.

5.4.3   *Method of global analysis*

The condition of EN 1993-1-1 for the use of first-order analysis is $\alpha_{cr} \geq 10$, where α_{cr} is the factor by which the design loading would have to be increased to cause elastic instability in a sway mode.

There is a well-known hand method of calculation of α_{cr} for simple frames involving s and c functions, which have been tabulated [50]. Computer programs are available for more complex frames. For beam-and-column plane frames in buildings, EN 1993-1-1 gives the approximation

$$\alpha_{cr} = (H_{Ed}/V_{Ed}) (h/\delta_{H,Ed}) \tag{5.8}$$

to be satisfied separately for each storey of height h. In this expression, V_{Ed} and H_{Ed} are the total vertical and horizontal reactions at the bottom of the storey, with forces $N\phi$ included in H_{Ed}, and $\delta_{H,Ed}$ is the change in lateral deflection over the height h.

It will be shown later that, for the frame in Fig. 5.1, the lateral stiffness of the floor slabs is so much greater than that of the columns that almost the whole of the lateral load is transferred to the core and end walls. These are stiff enough for $\delta_{H,Ed}$ to be very small, so α_{cr} for each storey far exceeds 10. Hence, first-order global analysis can be used for each frame. The shear walls require a separate check.

5.4.4 *First-order global analysis of braced frames*

5.4.4.1 *Actions*

This Section should be read with reference to Section 4.3, on global analysis of continuous beams, much of which is applicable. Braced frames do not have to be designed for horizontal actions, so the load cases are similar to those for beams. Imposed load is the leading variable action, and neither wind nor the $N\phi$ horizontal forces need be included.

No serviceability checks are normally required for braced frames, or for composite columns. The columns are designed using elastic global analysis and plastic section analysis; that is, as if their cross-sections were in Class 2. The method of construction, propped or unpropped, is therefore not relevant.

For most types of imposed load, the probability of the occurrence of the factored design load becomes less as the loaded area increases. EN 1991-1-1 recommends that the imposed load on a floor or roof with a loaded area of A m^2 may be reduced by a factor α_A, given by:

$$\alpha_A = 5\psi_0/7 + A_0/A \leq 1.0 \tag{5.9}$$

where $A_0 = 10.0$ m^2 and ψ_0 is the combination factor for the relevant type of imposed load. For the loading used in the example, $\psi_0 = 0.7$, so $\alpha_A = 1.0$ (no reduction) for loaded floor areas of less than 20 m^2.

The characteristic imposed load on a column that carries load from n storeys may be reduced by applying the recommended factor

$$\alpha_n = [2 + (n - 2)\psi_0]/n \tag{5.10}$$

For $\psi_0 = 0.7$, this gives $\alpha_n < 1$ for three or more storeys.

Maximum bending moments in columns in rigid-jointed frames occur when some of the nearby floors do not carry imposed load. For a column length AB in a frame with many similar storeys, the most adverse combination of axial load and bending moment is likely to occur when the imposed load is applied as in Fig. 5.7(a), for an external column, or Fig. 5.7(b), for an internal column. The bending-moment distributions for these columns are likely to be as shown.

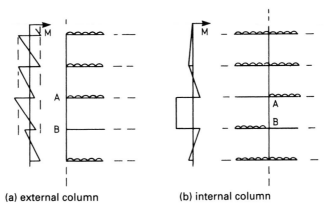

(a) external column   (b) internal column

Figure 5.7   Arrangements of imposed load, for column design

5.4.4.2 *Eccentricity of loading, for columns*

The use of 'nominally-pinned' beam-to-column joints reduces bending moments in the columns, with corresponding increases in the sagging moments in the beams. For the beams, it is on the safe side to assume that the moments in the connections are zero. If this were true, the load from each beam would be applied to the column at an eccentricity slightly greater than half the depth, h_a, of the steel column section (Fig. 5.2(a)) for major-axis connections.

An elastic analysis that modelled the real (non-zero) stiffness of the connection would give an equivalent eccentricity greater than this. The real behaviour is more complex. Initially, the end moments increase the tendency of each column length to buckle; but as the load increases, the column becomes less stiff, the end moments change sign, and the greater the stiffness of the connections, the more beneficial is restraint from the beams on the stability of the column.

Typically, British codes of practice for steel columns have allowed for this stabilising effect by modelling each storey-height column with an 'effective length' of between 70% and 85% of its actual length (e.g., $L_e/L \approx 0.7$), but have specified an 'equivalent eccentricity of loading', $e_c \approx 0.5h_a + 100$ mm, for the calculation of end moments. For the 206 × 204 UC section to be used here, it will be assumed that the load from a supported major-axis beam acts at 0.2 m from the axis of the column. This applies a bending moment to the column that is independent of its stiffness, and reduces the design span of the beam by 0.2 m at each joint.

In some other European countries, the practice has been to assume $L_e \approx L$, which makes buckling more critical, and $e_c = 0$, which eliminates these bending moments from columns. One justification for using $e_c = 0$ is that the bending moments in the beams are calculated using the span between column centres, rather than the smaller span between the centres of the 'pin' connections. Eurocodes 3 and 4 at present give no guidance on this subject.

In the following example, it will be assumed that $L_e = L$ and that the load from a nominally pinned connection acts at 100 mm from the face of the steel column section, so that

$$e_c = 0.5h_a + 100 \text{ mm}$$

The beam is assumed to be simply-supported at the pin, so its span is less than that to the column centre-line.

5.4.4.3 *Elastic global analysis*

This method of analysis is generally applicable to braced composite frames with rigid or nominally pinned joints. The flexural stiffness of hogging moment regions of beams is treated as in Section 4.3.2. For columns, concrete is assumed to be uncracked, and the stiffness of the longitudinal reinforcement is usually included, as it may not be negligible.

Bending moments in beams may be redistributed as in Section 4.3.2, but end moments found for composite columns may not be reduced, because there is insufficient knowledge of the rotation capacity of columns.

Where the beam-to-column joints are nominally pinned, as in the external columns in the example in Section 5.7, the bending moments in a column are easily found by moment distribution for that member alone.

5.4.4.4 *Rigid-plastic global analysis*

The use of this method for a braced frame is not excluded by EN 1994-1-1, but there are several conditions, which make it unattractive in practice. In addition to the conditions that apply to beams (Section 4.3.3), these include the following.

(1) All connections must be shown to have sufficient rotation capacity, or must be full-strength connections with $M_{j,Rd} \geq 1.2 M_{pl,Rd}$, as explained in Section 5.3.2.
(2) Unless verified otherwise, it should be assumed that composite columns do not have rotation capacity.

5.4.5 *Outline sequence for design of a composite braced frame*

This over-view is intended to provide an introduction to the subject; it is not comprehensive. Its scope is limited to regular multi-storey braced frames of the type used in the examples in Chapters 3, 4 and 5. It is assumed that the detailing will provide the required resistance to fire, and that the following decisions are made at the outset:

- number of storeys, storey heights, column positions, layout and spans of beams;
- use of floors to transfer lateral forces to bracing elements;
- type and location of bracing elements;
- imposed floor loadings and wind loading;
- type(s) of beam-to-column joints (pinned, semi-rigid, or rigid);
- nominal eccentricity for any nominally-pinned joints to columns;
- strengths of materials and densities of concretes to be used.

Ultimate limit states

(1)   Design the floor slabs (concrete or composite), spanning between the beams.

(2)   Find imposed-load reductions (if any) for the beams.

(3)   Do preliminary designs for beams, neglecting interaction with internal columns, as these have little influence, even if joints are rigid. Use of nominally-pinned joints to external columns enables their influence on beam design also to be ignored. Include the flexibility of any semi-rigid joints in analyses of continuous beams.

(4)   Find imposed-load reductions for columns, and do preliminary designs. Check that columns are not so slender that their imperfections should be included in global analysis.

(5)   Consider creep and cracking of concrete, and find elastic stiffnesses for all beams and columns.

(6)   Do elastic first-order global analyses for all frames, for gravity loads only, to find action effects in beams and columns. Moments in beams may be redistributed, within permitted limits. Neglect frame and member imperfections.

(7)   Check beam designs. Revise if necessary.

(8)   Increase bending moments in each column to allow for second-order effects within the column length and for column imperfections. Check column designs and revise if necessary.

(9)   Allow for frame imperfections by notional horizontal forces. Compare these with forces from wind, and decide whether, for horizontal loading, the leading variable action should be imposed gravity loading or wind loading.

(10)   Do preliminary designs for bracing elements, to find their stiffnesses.

(11)   Do elastic first-order global analyses for the complete structure, for horizontal loads plus appropriate gravity loads, to find lateral deflections. The arrangement of imposed loading should be that which gives maximum side-sway.

(12)   Repeat (11) with all beam–column intersections displaced laterally by the amounts found in step (11). Check that the increases in the action effects that govern design of members are all less than 10%. (This condition for the use of first-order global analyses is assumed to be satisfied.)

(13)   Check the design of the bracing elements, taking account of imperfections and second-order effects. (The example in this chapter does not include this.)

Design for serviceability limit states

Re-analyse the frames for unfactored vertical loads to check deflections and susceptibility to vibration. Detail reinforcement to control crack widths, as necessary.

5.5   Example: composite frame

5.5.1   *Data*

To enable previous calculations to be used, the structure to be designed has a composite slab floor that spans 4.0 m between two-span continuous composite beams with spans of 9.5 m. There are nine storeys, each with floor-to-floor height of 4.0 m, as shown in Fig. 5.1. The outer columns are assumed to be nominally pinned at ground level, and the internal columns to be nominally pinned at basement level, 4.0 m below the beam at ground level. For simplicity, it is assumed that the roof has the same loading and structure as the floors, though this would not be so in practice. The building stands alone, and the horizontal span of its floors between lateral restraints is 28 m, as explained earlier.

The materials and loadings are as used previously, Sections 3.2, 3.11 and 4.6, and the composite floor is as designed in Section 3.4. The two-span composite beams are as designed in Section 4.6, with nominally-pinned connections to the external columns (Section 5.10), except that they are not continuous over a central point support. There is instead a composite column at mid-length of each beam, to which each span is connected by a 'rigid' and 'full-strength' joint. These terms are defined in Section 5.3.2.

The only gravity loads additional to those carried by the beams are the weight of the columns and the external walls. The characteristic values are assumed to be as follows:

- for each column, $g_k = 3.0$ kN/m $= 12.0$ kN per storey
- for each external wall, $g_k = 60$ kN per bay per storey

The 60-kN load is for $4 \times 4 = 16$ m^2 of wall, which is assumed to be supported at each floor level by a beam spanning 4.0 m between adjacent columns.

The design ultimate gravity load per storey for each column is therefore the load from one main beam plus:

- for the internal column, $12 \times 1.35 = 16.2$ kN
- for an external column, $72 \times 1.35 = 97.2$ kN

$$(5.11)$$

The characteristic wind load is based on wind in a direction parallel to the longitudinal axes of the main beams. It causes pressure on the windward wall and suction (i.e., pressure below atmospheric) on the leeward wall. The sum of these two effects is assumed to be:

$$q_{k,wind} = 1.5 \text{ kN/m}^2 \text{ of windward wall} \qquad (5.12)$$

The effects of wind blowing along the building are not considered.

The properties of materials are as summarised in Section 4.6.1, except that the concrete in the composite columns is of normal density, with properties

$$f_{ck} = 25 \text{ N/mm}^2 \quad E_{cm} = 31.0 \text{ kN/mm}^2 \qquad (5.13)$$

The design initial side-sway of a frame such as DEF in Fig. 5.1 was found in Section 5.4.1 to be $\phi = 1/366$.

5.5.2 *Design action effects and load arrangements*

The whole of the design variable load for a typical frame is transferred to its three columns by the major-axis beams. Permanent loading is symmetrical about the plane of the frame, so the minor-axis bending moments applied to the columns are negligible. The additional gravity loads (Expressions 5.11) are assumed to cause no major-axis bending moments. These are caused in the external columns only by the loads from the major-axis beams.

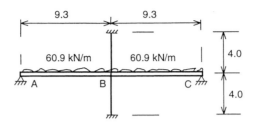

Figure 5.8 Beam loading for maximum hogging bending moment at B

Load arrangements

For each action effect in a member, the appropriate arrangement of imposed load is that which gives the most adverse value (usually, the highest value).

For hogging bending in a beam, and for axial force in columns, imposed load should be applied to both spans of every beam. From Table 4.4, the design loads are as shown in Fig. 5.8. The loaded floor area is $18.6 \times 4 = 72.4$ m^2 so, from Equation 5.9, a reduction factor $\alpha_A = 0.5 + 10/72.4 = 0.64$ could be applied. This is not done, for simplicity, so that earlier results can be used.

From symmetry, the loading causes no bending in the internal column. From Section 4.6.2, the bending moment in the beam at B, after 30% redistribution of moments, is

$$M_{Ed,B} = 461 \text{ kN m} \tag{5.14}$$

From equilibrium of length AB, the shear force in each beam at B is

$$V_{Ed,B} = 60.9 \times 9.3/2 + 461/9.3 = 333 \text{ kN} \tag{5.15}$$

To obtain the shear forces at the pins at A and C, each span is increased to 9.5 m, so that the loads on the external columns include the correct width of floor. Resolving vertically for span AB,

$$V_{A,Ed} = V_{C,Ed} = 60.9 \times 9.5 - 333 = 246 \text{ kN} \tag{5.16}$$

For maximum sagging bending moment in a beam, it is accurate enough to analyse the limited frame of Fig. 5.8 with the loading on one span reduced to 23.7 kN/m, the design dead load (Table 4.4).

For the columns, the arrangement of variable load shown in Fig. 5.9, with full variable load on all upper floors, will provide an adverse combination of axial force and single-curvature bending in both the column

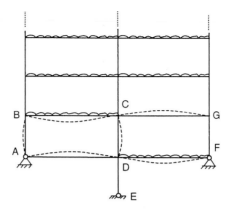

Figure 5.9 Arrangement of variable load for maximum bending in columns AB and CD

lengths AB and CD. It is assumed that column length DE can have an increased cross-section, if necessary.

The example is continued after the design methods for composite columns have been explained.

5.6 Simplified design method of EN 1994-1-1, for columns

5.6.1 *Introduction*

Background information for this Section is provided in Sections 5.1 and 5.2. Global analysis provides for each column length in a plane frame a design axial force, N_{Ed}, and applied end moments $M_{1,Ed}$ and $M_{2,Ed}$. By convention, M_1 is the greater of the two end moments, and they are both of the same sign where they cause single-curvature bending.

Initially, concrete-encased H- or I-sections are considered (Fig. 5.10(a) and (b)). Where methods for concrete-filled steel tubes (Fig. 5.10(c)) are different, this is explained in Section 5.6.7. The encased sections are assumed to have bi-axial symmetry, and to be uniform along each column length. Applied moments are resolved into the planes of major-axis and minor-axis bending of the column, and their symbols have additional subscripts (y and z, respectively) where necessary.

The two ends of a column length are assumed each to be connected to one or more beams and to be braced laterally at these points, distance L apart. The effective length of each column length is here assumed to be L, as explained in Section 5.4.4.2. Lateral loads on columns are assumed to be applied only at the ends of each column length.

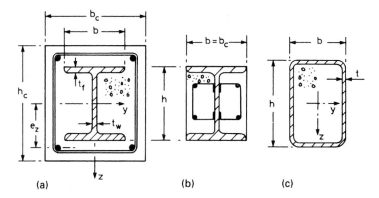

Figure 5.10   Typical cross-sections of composite columns

The methods explained below are applied separately for each plane of bending. It often happens that all significant bending occurs in one plane only. If this is minor-axis bending, no major-axis verification is needed. If it is major-axis bending, minor-axis buckling must be checked, as explained in Section 5.6.5.2, because of interaction between the axial load and the minor-axis imperfections.

These methods are different from column design to BS 5400:Part 5, and also from design of steel columns to EN 1993-1-1.

5.6.2   *Fire resistance, and detailing rules*

Before doing calculations based on an assumed cross-section for a composite column, it is wise to check that the section satisfies relevant limits to its dimensions.

The resistance to fire of a concrete-encased I-section column is determined by the thickness of the concrete cover to the steel section and the reinforcement. For a 90-minute period of resistance and a cross-section with dimensions h_c and b_c of at least 250 mm (for example), the limits to cover given by EN 1994-1-2 are 40 mm to the steel section and 20 mm to the reinforcement.

The rules of EN 1992-1-1 for minimum cover and reinforcement, and for maximum and minimum spacing of bars, should be followed. These ensure resistance to corrosion, safe transmission of bond forces, and avoidance of spalling of concrete and buckling of longitudinal bars. The ratio of area of reinforcement to area of concrete allowed for in calculating resistances should satisfy

$$0.003 \leq A_s/A_c \leq 0.06 \tag{5.17}$$

The upper limit is to ensure that the bars are not too congested at overlaps.

The thickness of concrete cover to the steel section that may be used in calculations has upper limits $c_y = 0.4b$, $c_z = 0.3h$. These relate to the proportions of columns for which this design method has been validated. The *steel contribution ratio* δ and the *slenderness* $\overline{\lambda}$ (Section 5.6.3.1) are limited for the same reason.

The steel contribution ratio is defined by

$$\delta = A_a f_{yd}/N_{pl,Rd} \tag{5.18}$$

where f_{yd} is the design yield strength of the structural steel, with the condition

$$0.2 \le \delta \le 0.9$$

If $\delta < 0.2$, the column should be treated as reinforced concrete; and if $\delta > 0.9$, as structural steel. The term $A_a f_{yd}$ is the contribution of the structural steel section to the plastic resistance $N_{pl,Rd}$, given by Equation 5.24.

5.6.3 *Properties of column lengths*

The characteristic elastic flexural stiffness of a column cross-section about a principal axis (y or z) is the sum of contributions from the structural steel (subscript a), the reinforcement (subscript s) and the concrete (subscript c), and so has the format:

$$(EI)_{eff} = E_a I_a + E_s I_s + K_c E_{c,eff} I_c \tag{5.19}$$

where E is the elastic modulus of the material and I the second moment of area of the relevant cross-section.

The elastic critical normal force is found from

$$N_{cr} = \pi^2 (EI)_{eff}/L^2 \tag{5.20}$$

where L should be taken as the length between the lateral restraints in the plane of buckling considered. The 'concrete' term in Equation 5.19 is based on calibration of results from this method against test data. It was found that $K_c = 0.6$ and that creep should be allowed for by reducing the mean short-term elastic modulus for concrete, E_{cm}, as follows:

$$E_{c,eff} = E_{cm}/[1 + (N_{G,Ed}/N_{Ed})\varphi_t] \tag{5.21}$$

where the symbols are defined after Equation 5.7.

As N_{cr} does not depend on strengths of materials, no partial factors are involved. It is treated as a 'characteristic' value. The use of the characteristic value of EI, denoted $(EI)_{eff}$, is therefore appropriate. A lower 'design' value is also needed, for analysis of second-order effects within a column length. This is given in EN 1994-1-1 as

$$(EI)_{eff,II} = 0.9(E_a I_a + E_s I_s + 0.5 E_{c,eff} I_c) \tag{5.22}$$

with $E_{c,eff}$ from Equation 5.21. The factor 0.5 allows for cracking, and the 0.9 is based on calibration work.

5.6.3.1 *Relative slenderness*

The non-dimensional relative slenderness of a column length for buckling about a particular axis is defined by

$$\bar{\lambda} = \sqrt{(N_{pl,Rk}/N_{cr})} \tag{5.23}$$

The design resistance to axial load of a straight column too short to buckle, known as the 'squash load', is given by

$$N_{pl,Rd} = A_a f_{yd} + A_s f_{sd} + 0.85 A_c f_{cd} \tag{5.24}$$

where the design strengths of the materials are:

- for structural steel, $\qquad$ $f_{yd} = f_y/\gamma_A$ (not f_{yk} because f_y is a nominal value)
- for reinforcement, $\qquad$ $f_{sd} = f_{sk}/\gamma_S$
- for concrete in compression, $f_{cd} = f_{ck}/\gamma_C$

and the γs are the usual partial factors for ultimate limit states. The area A_c is conveniently calculated from

$$A_c = b_c h_c - A_a - A_s \tag{5.25}$$

For calculating $\bar{\lambda}$, $N_{pl,Rd}$ is replaced by the characteristic squash load,

$$N_{pl,Rk} = A_a f_y + A_s f_{sk} + 0.85 A_c f_{ck} \tag{5.26}$$

because N_{cr} is a characteristic value.

The following method of column design, from EN 1994-1-1, is limited to column lengths with $\bar{\lambda} \leq 2$. This limit is rarely exceeded in practice.

5.6.4 *Resistance of a cross-section to combined compression and uni-axial bending*

Design for a combination of load along the x-axis and bending about the y- or z-axis is based on an interaction curve between resistance to compression, N_{Rd}, and resistance to bending about the relevant axis, M_{Rd}. The method is explained with reference to Fig. 5.11. The plastic resistance $N_{pl,Rd}$ is given above.

The complexity of hand methods of calculation for $M_{pl,Rd}$ and other points on the curve has been a disincentive to the use of composite columns. It is quite easy to prepare a spreadsheet to do this. However, it should be noted that when a rolled I- or H-section is represented by three rectangles, as in the algebra given in Reference 17 and outlined below, results will differ slightly from those by hand calculation, unless the corner fillets are allowed for.

The assumptions are those used for calculating $M_{pl,Rd}$ for beams: rectangular stress blocks with structural steel at a stress $\pm f_{yd}$, reinforcement at $\pm f_{sd}$, and concrete at $0.85f_{cd}$ in compression or cracked in tension. Full shear connection is assumed.

The complexity appears in the algebra. For major-axis bending of the section shown in Fig. 5.10(a), there are five possible locations of the plastic neutral axis, each leading to different expressions for N_{Rd} and M_{Rd}. A practicable method is to guess a position for the neutral axis, and calculate N_{Rd} by summing the forces in the stress blocks, and M_{Rd} by taking moments of these forces about the centroid of the uncracked section. This gives one point on Fig. 5.11. Other points, and hence the curve, are found by repeating the process.

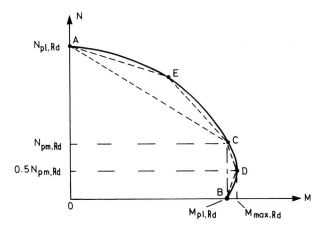

Figure 5.11 Polygonal approximation to *M–N* interaction curve

The simplification made in EN 1994-1-1 is to replace the curve by a polygonal diagram, AECDB in Fig. 5.11. An ingenious and fairly simple method of calculating the co-ordinates of points B, C and D was given in the ENV version of EN 1994-1-1, and is now explained in an Appendix of Reference 17. It is used in Section 5.7.2.

For major-axis bending of encased I-sections, AC may be taken as a straight line, but for other situations an intermediate point, E, should be found, as line AC can be too conservative. For point E, the first guessed neutral-axis position is usually good enough. A similar method is used for the interaction polygon for axial load and minor-axis bending.

Transverse shear force may be assumed to be resisted by the steel section alone. The design method for moment-shear interaction in beams (Section 4.2.2) may be used. In columns, V_{Ed} is usually less than $0.5V_{pl,Rd}$, and then no reduction in bending resistance need be made. None is assumed here.

5.6.5 *Verification of a column length*

5.6.5.1 *Design action effects, for uni-axial bending*

It is assumed that the interaction curve or polygon, Fig. 5.11, has been determined, and that the design axial force N_{Ed} and the end moments $M_{1,Ed}$ and $M_{2,Ed}$ have been found by global analysis. It is rare for a vertical column length in a building to be subjected to significant transverse load within its length, and none is assumed here.

If, as is likely, member imperfections (see Section 5.4.1) were omitted from the global analysis of the frame, the initial bow, of amplitude e_0, is allowed for now. Its first-order effect is to increase the bending moment at mid-length of the column by $N_{Ed}e_0$.

The condition in EN 1994-1-1 for neglect of second-order effects is

$$N_{cr,eff} \geq 10N_{Ed} \tag{5.27}$$

where $N_{cr,eff}$ is found from Equation 5.20 with $(EI)_{eff}$ replaced by $(EI)_{eff,II}$ from Equation 5.22. If this condition applies, the design bending moment M_{Ed} for the column is the greatest value given by the curve in Fig. 5.12(a). Otherwise, a second-order analysis is required, or the following simplified methods from EN 1994-1-1 should be used.

Second-order effects in a column length
Second-order effects of the end moments and from the $N_{Ed}e_0$ moment are found separately, and can be superimposed. This is possible because they both result from the same axial force. They are always added, because the imperfection e_0 can occur in any lateral direction. Subscripts 'end' and 'imp' are now used, respectively, for these two sets of moments.

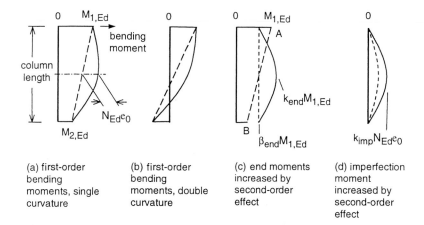

(a) first-order bending moments, single curvature

(b) first-order bending moments, double curvature

(c) end moments increased by second-order effect

(d) imperfection moment increased by second-order effect

Figure 5.12 First-order and second-order bending moments in a column length

The greatest first-order bending moment is multiplied by a factor k_{end} given by

$$k = \beta/[1 - (N_{Ed}/N_{cr,eff})] \tag{5.28}$$

where

$$\beta = 0.66 + 0.44(M_2/M_1) \geq 0.44 \tag{5.29}$$

with $N_{cr,eff}$ as above. The coefficient β allows for the more adverse effect of single-curvature bending than of double-curvature bending, for which M_2/M_1 is negative.

In EN 1994-1-1, Equation 5.28 appears with the further condition $k \geq 1.0$. It is over-conservative to apply this when combining two sets of second-order effects, and β_{end} is often such that $k_{end} < 1$. This value need not be increased to 1.0. The first-order end moments, for example AB in Fig. 5.12(c), are replaced by an equivalent uniform moment $\beta_{end}M_{1,Ed}$, which is increased to $k_{end}M_{1,Ed}$ at mid-length to allow for second-order effects, as shown. This always exceeds $\beta_{end}M_{1,Ed}$, from Equation 5.28.

For the bending moment from the member imperfection, EN 1994-1-1 specifies $\beta_{imp} = 1.0$ so, from Equation 5.28, k_{imp} always exceeds 1.0, Fig. 5.12(d).

The design bending moment for the column length is usually

$$M_{Ed} = k_{end}M_{1,Ed} + k_{imp}N_{Ed}e_0 \tag{5.30}$$

but is $M_{1,Ed}$, if greater.

Verification, for uni-axial bending

The column is strong enough if its cross-section can resist the combination of M_{Ed} with N_{Ed}. The bending resistance M_{Rd} in the presence of axial compression N_{Ed} is found from the interaction diagram, explained in Section 5.6.4.

A correction is required for the unconservative assumption that the rectangular stress block for concrete extends to the plastic neutral axis (Section 3.5.3.1). It is made by reducing the bending resistance, so that the verification condition is

$$M_{Ed} \leq \alpha_M M_{Rd} \tag{5.31}$$

where $\alpha_M = 0.9$ for steel grades up to S355, for which $f_y = 355$ N/mm² for sections of thickness up to 40 mm. It is reduced to 0.8 for stronger steels, to take account of the adverse effect of their higher yield strain on the load at which concrete begins to crush.

5.6.5.2 *Bi-axial bending*

It has to be decided in which plane of bending failure is expected to occur. This is usually obvious. The bending moment $N_{Ed}e_0$ is included only for this plane. The axial load N_{Ed} and the maximum design bending moments about both axes, $M_{y,Ed}$ and $M_{z,Ed}$, are found, as in Section 5.6.5.1. The verification consists of checking Expression 5.31 separately for each axis and, in addition, satisfying Expression 5.32:

$$M_{y,Ed}/M_{y,Rd} + M_{z,Ed}/M_{z,Rd} \leq 1.0 \tag{5.32}$$

5.6.6 *Transverse and longitudinal shear*

For applied end moments M_1 and M_2, as defined in Section 5.6.1, the transverse shear in a column length is $(M_1 - M_2)/L$. An estimate can be made of the longitudinal shear stress at the interface between steel and concrete, by elastic analysis of the uncracked composite section. This is rarely necessary in multi-storey structures, where these stresses are usually very low.

Higher stresses may occur near joints at a floor level where the axial load added to the column is a high proportion of the total axial load. Load added after the column has become composite, N_{Ed} say, is assumed to be transferred initially to the steel section, of area A_a. It is then shared between the steel section and its encasement on a transformed area basis:

$$N_{Ed,c} = N_{Ed}(1 - A_a/A) \tag{5.33}$$

where $N_{Ed,c}$ is the force that causes shear at the surface of the steel section and A is the transformed area of the column in 'steel' units. There must be a 'clearly defined load path . . . that does not involve an amount of slip at this interface that would invalidate the assumptions made in design' (from EN 1994-1-1).

There is no well-established method for calculating longitudinal shear stress at the surface of the steel section, τ_{Ed}. Design is usually based on mean values, found by dividing the force by the perimeter of the section, u_a, and an assumed 'load introduction length', ℓ_V:

$$\tau_{Ed} = N_{Ed,c}/u_a\ell_V \qquad (5.34)$$

It is recommended in EN 1994-1-1 that ℓ_V should not exceed the least of $2b_c$, $2h_c$ (Fig. 5.10) and $L/3$, where L is the column length.

Design shear strengths τ_{Rd} due to bond and friction are given in EN 1994-1-1 for several situations. For completely encased sections,

$$\tau_{Rd} = 0.3 \text{ N/mm}^2 \qquad (5.35)$$

This is a low value, to take account of the approximate nature of τ_{Ed}.

Where τ_{Ed} is less than τ_{Rd}, no account need be taken of the further transfer of force by shear between steel and concrete as failure is approached. The best protection against local failure is provided by the transverse reinforcement (links) which are required by EN 1992-1-1 to be more closely spaced near beam–column intersections than elsewhere.

In regions where τ_{Rd} is exceeded, shear connectors should be provided for the whole of the shear. These are best attached to the web of a steel H- or I-section, because their resistance is enhanced by the confinement provided by the steel flanges. Design rules are given in EN 1994-1-1.

5.6.7 *Concrete-filled steel tubes*

A typical cross-section of a column of this type is shown in Fig. 5.10(c). To avoid local buckling of the steel, slendernesses of the walls must satisfy

$$h/t \leq 52\varepsilon \qquad (5.36)$$

where

$$\varepsilon = (235/f_y)^{0.5}$$

and f_y is the yield strength in N/mm^2 units. For concrete-filled circular hollow sections of diameter d the limit is more generous:

$$d/t \leq 90\varepsilon^2 \qquad (5.37)$$

Design is essentially as for encased H-sections, except that in calculating the squash load $N_{pl.Rd}$, account is taken of the higher resistance of the concrete, caused by lateral restraint from the steel tube, as follows.

The factor 0.85 in Equations 5.24 and 5.26 is replaced by 1.0. Also, for circular sections only, f_{cd} is increased to an extent that depends on the ratios t/d, f_y/f_{ck}, $\bar{\lambda}$ and $M_{Ed}/(N_{Ed}d)$, provided that the relative slenderness $\bar{\lambda} \leq 0.5$.

For a circular section, there is also a reduction in the effective yield strength of the steel wall used in calculating $N_{pl.Rd}$, to take account of the circumferential tensile stress in the wall. This stress provides restraint to lateral expansion of the concrete caused by the axial load on the column. These rules are based on extensive testing.

5.7 Example: external column

5.7.1 *Action effects*

The use of nominally-pinned joints and a braced frame enables the design of an external column to be completed without further global analysis of the frame of Fig. 5.1. From Section 5.5, the design ultimate shear force from a fully-loaded beam is 246 kN. The size of the external column is governed by the length 0–1 in Fig. 5.13(a). This supports load from nine floors, so from Equation 5.10 the reduction factor for imposed load, with $\psi_0 = 0.7$ as before, is

$$\alpha_n = (2 + 7 \times 0.7)/9 = 0.767$$

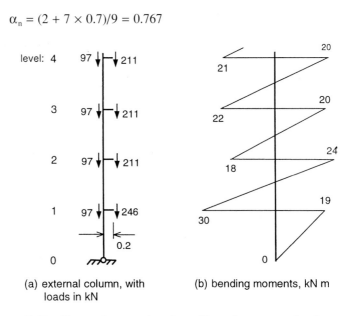

(a) external column, with
loads in kN

(b) bending moments, kN m

Figure 5.13   Dimensions and action effects for external column

From Table 4.4, the variable load provides $37.2/60.9 = 61\%$ of the shear force. Hence, the shear force at each pin joint is

$$V_{Ed} = 246 \times 0.61 \times 0.767 + 246 \times 0.39 = 115 + 96 = \textbf{211 kN}$$
(5.38)

It is assumed that the steel section for the column will be from the 203×203 UC serial size, so the eccentricity at the pin joint is $0.203/2 + 0.1 = 0.20$ m, and the major-axis bending moment applied to the column at each loaded floor level is

$$M_{Ed} = 211 \times 0.2 = 42.2 \text{ kN m}$$
(5.39)

The bending moment in length 0–1 is determined mainly by the load from level 1, so factor α_n is taken as 1.0 at that level, giving the applied moments and shears shown in Fig. 5.13(a). The bending moments in the lower part of the column, found by moment distribution, are shown in Fig. 5.13(b).

For minor-axis bending, all the loading is permanent, and equal on the two sides of the column, so bending moment arises only from the initial bow of the member.

Including the permanent load from Equation 5.11, the axial load for length 0–1 is:

$$N_{Ed} = 9 \times 97 + 8 \times 211 + 246 = \textbf{2807 kN}$$
(5.40)

5.7.2 *Properties of the cross-section, and y-axis slenderness*

A cross-section for the column must now be assumed, and is shown in Fig. 5.14. Applying the usual partial factors to the properties of the

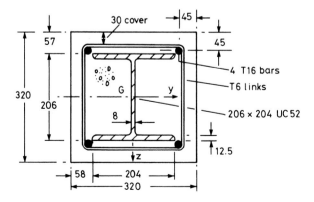

Figure 5.14 Assumed cross-section for external column length 0–1

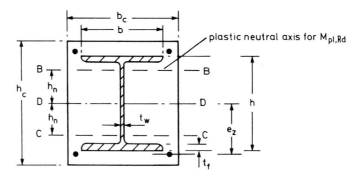

Figure 5.15   Plastic neutral axes for encased I-section

The plastic section moduli for the three materials, assuming that concrete is as strong in tension as in compression, are:

$$10^{-6}W_{pa} = 0.568 \text{ mm}^3 \text{ (from tables)}$$

$$10^{-6}W_{ps} = A_s e_z = 0.804 \times 0.115 = 0.0925 \text{ mm}^3$$

$$10^{-6}W_{pc} = b_c h_c^2/4 - 0.568 - 0.0925 = 7.53 \text{ mm}^3$$

For rectangular stress blocks with stresses $f_{yd} = \pm355 \text{ N/mm}^2$ in steel, $f_{sd} = \pm435 \text{ N/mm}^2$ in reinforcement and $0.85f_{cd} = 14.2 \text{ N/mm}^2$ in concrete, in compression only, it is found from longitudinal equilibrium with $N_{Ed} = 0$ that $h_n = 67$ mm.

Plastic section moduli for the region of depth $2h_n$ between lines B–B and C–C in Fig. 5.15 are now found:

$$10^{-6}W_{pa,n} = t_w h_n^2 = 8 \times 0.067^2 = 0.036 \text{ mm}^3$$

$$10^{-6}W_{pc,n} = (b_c - t_w)h_n^2 = (320 - 8) \times 0.067^2 = 1.40 \text{ mm}^3$$

At point D on the interaction polygon of Fig. 5.11, the neutral axis is line D–D in Fig. 5.15. The longitudinal forces in the steel section and the reinforcement sum to zero, from symmetry, so the axial compression is $N_{pm,Rd}/2$, where

$$N_{pm,Rd} = 0.85A_c f_{cd} \tag{5.46}$$

Hence,

$$\mathbf{0.5}N_{pm,Rd} = 0.5 \times 94.95 \times 14.2 = \mathbf{673 \text{ kN}}$$

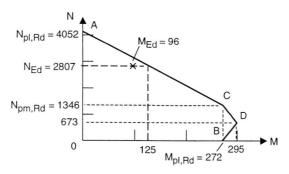

Figure 5.16 Interaction diagram for major-axis bending of external column

The bending resistance at point D, with W_{pc} halved to allow for cracking, is

$$M_{max,Rd} = W_{pa}f_{yd} + W_{ps}f_{sd} + 0.85W_{pc}f_{cd}/2$$

$$= 0.568 \times 355 + 0.0925 \times 435 + 7.53 \times 14.2/2$$

$$= 295 \text{ kN m} \tag{5.47}$$

When the plastic neutral axis moves from D–D to C–C, the axial compression changes from $N_{pm,Rd}/2$ to $N_{pm,Rd}$, because the changes in axial force are of the same size (but of opposite sign) as when it moves from D–D to B–B.

When the plastic neutral axis moves from B–B to C–C, the resultant of all the changes in axial force passes through G (from symmetry), so that the bending resistances at points B and C are the same, and are

$$M_{pl,Rd} = M_{max,Rd} - W_{pa,n}f_{yd} - W_{pc,n}f_{cd}/2$$

$$= 295 - 0.036 \times 355 - 1.4 \times 14.2/2$$

$$= 272 \text{ kN m} \tag{5.48}$$

The axial force at point C is

$$N_{pm,Rd} = 1346 \text{ kN} \tag{5.49}$$

so the position of line AC in Fig. 5.11 is as shown in Fig. 5.16.

5.7.3 *Resistance of the column length, for major-axis bending*

The design axial compression is $N_{Ed} = 2807$ kN from Equation 5.40; and from Equation 5.44:

$$N_{cr,eff} = \pi^2 \times 15.8 \times 1000/4^2 = 9750 \text{ kN}$$

The condition $N_{\text{cr,eff}} \geq 10 N_{\text{Ed}}$ (Equation 5.27) is not satisfied, so second-order effects must be allowed for.

From Fig. 5.13(b),

$$M_{1,\text{Ed}} = 19 \text{ kN m} \quad \text{and} \quad M_{2,\text{Ed}} = 0$$

From Equation 5.29,

$$\beta_{\text{end}} = 0.66$$

From Equation 5.28,

$$k_{\text{end}} = 0.66/(1 - 2807/9750) = 0.66 \times 1.404 = 0.93$$

For the initial bow $e_0 = 20$ mm (Section 5.4.1), $\beta_{\text{imp}} = 1.0$, and

$$N_{\text{Ed}} e_0 = 2807 \times 0.02 = 56 \text{ kN m}$$

From Equation 5.28,

$$k_{\text{imp}} = 1.404$$

From Equation 5.30, and as shown in Fig. 5.17,

$$M_{\text{y,Ed}} = 0.93 \times 19 + 1.4 \times 56 = 18 + 78 = \mathbf{96 \text{ kN m}} \tag{5.50}$$

From Fig. 5.16,

$$M_{\text{y,Rd}} = 125 \text{ kN m} \tag{5.51}$$

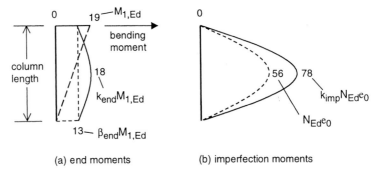

(a) end moments   (b) imperfection moments

Figure 5.17 Major-axis bending-moment diagrams for external column

From Equation 5.31,

$$\alpha_M M_{y,Rd} = 0.9 \times 125 = \textbf{113 kN m} \tag{5.52}$$

This exceeds $M_{y,Ed}$, so this column length has sufficient major-axis resistance. It can be shown that although the length above has a higher end moment, 30 kN m, it also is strong enough, because it is in double-curvature bending.

5.7.4 *Resistance of the column length, for minor-axis bending*

The margin of resistance to major-axis bending (above) is quite low, so two more T16 reinforcing bars were added to the cross-section, as shown in Fig. 5.18(a), to increase minor-axis resistance.

The cross-sectional areas given in Section 5.7.2 now become:

$$A_a = 6640 \text{ mm}^2 \quad A_s = 1206 \text{ mm}^2 \quad A_c = 94\,550 \text{ mm}^2$$

and $N_{pl,Rd}$ is increased from 4052 kN to 4224 kN.

The second moments of area are:

for the steel section, from tables, $10^{-6}I_a = 17.7 \text{ mm}^4$
for the reinforcement, $10^{-6}I_s = 1206 \times 0.115^2 = 15.9 \text{ mm}^4$
for the concrete, $10^{-6}I_c = 320^2 \times 0.32^2/12 - 17.7 - 15.9 = 840 \text{ mm}^4$

With $E_{c,eff} = 10.8 \text{ kN/mm}^2$ (Section 5.7.2), and from Equation 5.22,

$$10^{-12}(EI)_{eff,II} = 0.9(0.21 \times 17.7 + 0.20 \times 15.9 + 0.5 \times 0.0108 \times 840)$$

$$= 10.3 \text{ N mm}^2$$

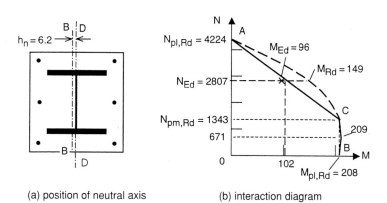

(a) position of neutral axis   (b) interaction diagram

Figure 5.18 Interaction diagram for minor-axis bending of external column

Using this value in Equation 5.20,

$$N_{cr,eff} = \pi^2 \times 10.3 \times 1000/4^2 = 6353 \text{ kN}$$

With $\beta_{imp} = 1.0$ as before, and from Equation 5.28,

$$k_{imp} = 1/(1 - 2807/6353) = 1.79$$

From EN 1994-1-1, the initial bow for the minor axis is $L/150 = 27$ mm so, from Equation 5.30,

$$M_{z,Ed} = 1.79 \times 2807 \times 0.027 = \textbf{136 kN m} \tag{5.53}$$

Interaction diagram for minor-axis bending

The method of calculation is similar to that used in Section 5.7.2. The plastic neutral axis for pure bending usually intersects the steel flanges, but not the web, and was found for this cross-section to be as shown in Fig. 5.18(a), with $h_n = 6.2$ mm.

The required plastic section moduli are:

$$10^{-6}W_{pa} = 0.263 \text{ mm}^3 \text{ (from tables)}$$

$$10^{-6}W_{ps} = 1.206 \times 0.115 = 0.139 \text{ mm}^3$$

$$10^{-6}W_{pc} = 3.2^3/4 - 0.263 - 0.139 = 7.79 \text{ mm}^3$$

$$10^{-6}W_{pa,n} = 0.004 \text{ mm}^3$$

$$10^{-6}W_{pc,n} = 0.010 \text{ mm}^3$$

As in Equation 5.47,

$$M_{max,Rd} = 0.263 \times 355 + 0.139 \times 435 + 7.79 \times 14.2/2 = 209 \text{ kN m}$$

As in Equation 5.48,

$$M_{pl,Rd} = 209 - 0.004 \times 355 - 0.010 \times 14.2/2 = \textbf{208 kN m}$$

From Equation 5.46,

$$N_{pm,Rd} = 94.55 \times 14.2 = 1343 \text{ kN}$$

The interaction polygon, shown in Fig. 5.18(b), gives

$$M_{z,Rd} = \textbf{102 kN m} \tag{5.54}$$

From Equation 5.31,

$$\alpha_M M_{z,Rd} = 0.9 \times 102 = \textbf{92 kN m} \tag{5.55}$$

This is less than $M_{z,Ed}$, so this column length is found to be too weak in minor-axis bending, when the polygonal approximation for curve AC in Fig. 5.18(b) is used.

Using the computed curve, $M_{z,Rd}$ is found to be 149 kN m, which appears to be sufficient; but resistance to bi-axial bending must be checked.

Bi-axial bending

For the bi-axial check to Expression 5.32, the initial bow is assumed to be in the more adverse plane, so either $M_{y,Ed}$ or $M_{z,Ed}$ is reduced. Here, $N_{Ed}e_0$ is greater for minor-axis bending, so $M_{y,Ed}$ is reduced from 96 kN m to 18 kN m, Fig. 5.17. From Equations 5.51 and 5.53, and using $M_{z,Rd} = 149$ kN m,

$$18/125 + 136/149 = 0.14 + 0.91 = 1.05$$

but should not exceed 1.0. To satisfy this check, $M_{z,Rd}$ would have to be increased to $136/0.86 = 158$ kN m. This could be done by providing more reinforcement.

5.7.5 *Checks on shear*

These checks are described in Section 5.6.6. The major-axis design transverse shear is greatest in column length 1–2, and is

$$V_{Ed} = (M_1 - M_2)/L = (24 + 30)/4 = 13.5 \text{ kN}$$

This is obviously negligible; $V_{pl,Rd}$ for the web of the steel section is over 300 kN.

From Fig. 5.13(b), the total vertical load applied to the column at level 1 is $246 + 97$ kN, but the self-weight of the column can be deducted, giving $N_{Ed} = 327$ kN. Any load transferred to the concrete encasement by direct bearing of the three steel beams connected to the column at this floor level is conservatively neglected. Creep reduces the load transfer, so the short-term modular ratio, $n_0 = 10.1$ is used for the transformed area of the column cross-section, A. With areas from Section 5.7.4,

$$10^{-3}A = 6.64 + 1.206 + 94.55/10.1 = 17.2 \text{ mm}^2$$

From Equation 5.33,

$$N_{Ed,c} = 327(1 - 6.64/17.2) = 201 \text{ kN}$$

The perimeter of the steel section is $u_a = 1140$ mm. Assuming a transmission length of 640 mm, twice the least lateral dimension, Equation 5.34 gives

$$\tau_{Ed} = N_{Ed,c}/u_a\ell_V = 201/(1.14 \times 640) = 0.27 \text{ N/mm}^2$$

This is less than τ_{Rd}, Equation 5.35, so local bond stress is not excessive, and shear connection is not required.

This completes the validation for this column length, provided that analysis for lateral loading (Section 5.9) confirms the assumption that it is all transferred by the floors to the end walls and the central core.

5.8   Example (continued): internal column

A typical internal column between level 0 and level 1 is now designed, for the arrangement of variable loading shown in Fig. 5.9. Full permanent load acts on all of the beams. Variable load acts on all beams at levels 2 to 9, but not on beams AD and CG, as this increases the single-curvature bending moment in length CD of the column. Any rotational restraint from point E, at basement level, is neglected.

In Section 5.7.1, the live-load reduction factor was found to be $\alpha_n = 0.767$. This reduces the design ultimate load on each beam from 60.9 kN/m (Table 4.4) to 52.2 kN/m. This reduction is not made for the beams that cause bending in the column, so the beam loadings for the global analysis are as shown in Fig. 5.19.

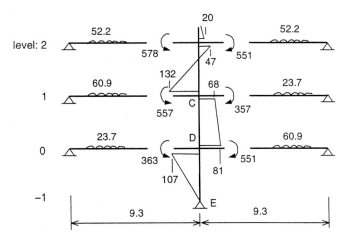

Figure 5.19 Loadings and major-axis bending moments for an internal column

Global analysis

It is assumed that the major-axis joints between the beams and the internal column are rigid and full-strength, and that the frame is braced against sidesway. The uncracked unreinforced second moment of area of the beam, with $n = 20.2$, is given in Table 4.5 as $I = 636 \times 10^6$ mm^4, so

$$(EI)_{\text{beam}} = 636 \times 210 \times 10^6 = 1.34 \times 10^{11} \text{ kN mm}^2 \qquad (5.56)$$

The column was at first assumed to have a 254 × 254 UC73 steel section, encased to 400 mm square, and with longitudinal reinforcement of six T20 bars. To obtain its stiffness, an effective modulus for the concrete is required. A low value, relative to that for the beams, reduces the bending moments in the column, so the creep coefficient $\varphi = 3.0$, used in Section 5.7.2 for resistance, may be too high. The design is relatively insensitive to this assumption, so the value $n = 2n_0$ is now used for the whole loading, as for the beams. For the Grade C25/30 normal-density concrete used, $E_{\text{cm}} = 31$ kN/mm^2, so $n = 2 \times 210/31 = 13.6$. This leads to

$$(EI)_{\text{column}} = 0.60 \times 10^{11} \text{ kN mm}^2 \qquad (5.57)$$

The beam members in Fig. 5.19 are over twice as long as the column members, so the stiffnesses $(EI)/L$ of beams and columns at nodes C, D, etc., are similar. Moment distribution for this limited frame gives the bending moments shown in Fig. 5.19, plotted on the tension side of each member.

For the beams, from Section 4.2.1.2, the bending resistance at the internal column is

$$M_{\text{pl,Rd}} = 510 \text{ kN m}$$

Cracking and inelastic behaviour are assumed to reduce the beam moments that exceed 510 kN m to that value, by redistribution of moments to mid-span, without altering the column moments. This enables the shear force in each beam at nodes C, D, etc., to be found, and hence, the total axial load in the column just above node D, including its weight. The result is

$$N_{\text{Ed}} = 5401 \text{ kN} \qquad (5.58)$$

Resistance of an internal column

Approximate calculations then showed that the initial column cross-section was too weak. The larger doubly-symmetric cross-section shown in Fig. 5.20 was assumed. Its resistance was checked by the methods used in Section 5.7 for external columns, taking account of the single-curvature

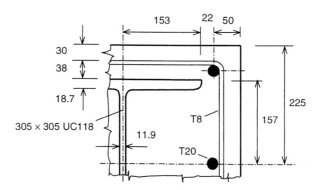

Figure 5.20   Part cross-section of revised internal column length 0–1

bending moments $M_{y,Ed,1} = 81$ kN m and $M_{y,Ed,2} = 68$ kN m (Fig. 5.19), the $N_{Ed}e_0$ moments, and the second-order moments about both axes; and checking uni-axial and bi-axial bending.

Major-axis bending was found to be the most critical, with $M_{y,Ed} = 216$ kN m and $0.9M_{y,Rd} = 443$ kN m, when $N_{Ed} = 5401$ kN. The margins on bending resistance so found were assumed to be sufficient to cover the increase in the end moments caused by the change in column cross-section, and the small increase in N_{Ed} from the heavier column.

Comment on column design
It is evident, above, that response to the uncertainties of loading, cracking, creep and inelastic behaviour involve some judgement and approximation. A small increase in the cross-section of a column reduces its slenderness (and hence, the secondary bending moments) as well as increasing all its resistances (N_{Rd}, $M_{y,Rd}$, etc.); so there is little saving in cost from seeking an 'only-just-adequate' design.

5.9   Example (continued): design for horizontal forces

As explained in Section 5.1, horizontal loads in the plane of a typical frame, such as DEF in Fig. 5.1(a), are transferred by the floor slabs to a central core and to two shear walls at the ends of the building, Fig. 5.21. It will be shown by approximate calculation that the system is so stiff, and the relevant stresses are so low, that rigorous verification is unnecessary.

It is shown in Section 5.4.1 that allowance for the frame imperfections is made by applying at each floor level of each frame a notional horizontal force $H_{Ed} = (G + Q)/366$, where G and Q are the total design ultimate dead and imposed loads for the relevant storey.

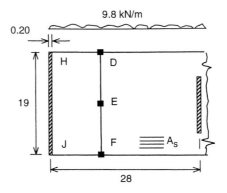

Figure 5.21 Part plan of typical floor slab

From EN 1994-1-1, second-order effects may be neglected in the global analysis if the deformations from first-order global analysis increase the relevant internal action effects by less than 10%. It will be found that this exemption applies.

It is assumed that the concrete above the profiled sheeting in each floor slab acts as a reinforced concrete beam of breadth 80 mm (its thickness) and depth 19 m, spanning 28 m. For simplicity, this span is assumed to be simply-supported. The lateral stiffnesses of these deep 'beams' and of the shear walls are so much higher than that of each frame, such as DEF in Fig. 5.21, that the presence of the frames can be ignored.

Design loadings, ultimate limit state
The force H_{Ed} is greatest when live load is applied to all floors, so the reduction factor $\alpha_n = 0.767$ (Section 5.7.1) is applicable. From Table 4.4, the imposed beam loading is

$0.767 \times 37.2 = 28.5$ kN/m

Dead loads are as in Section 5.5.1, except that the design weight of a 4-m length of internal column has increased from 16.2 kN to 30.0 kN.

The total permanent load per storey from a 4-m length of the building is

$G = 23.7 \times 19 + 2 \times 97.2 + 30 = 675$ kN

The imposed load is

$Q = 28.5 \times 19 = 542$ kN

From Section 5.5.1, with $\gamma_F = 1.5$, the design wind load is

$$W = 1.5 \times 1.5 \times 4 \times 4 = 36 \text{ kN}$$

These values are such that wind should be taken as the leading variable action. For imposed load, the combination factor $\psi_0 = 0.7$, and $\phi = 1/366$, so

$$H_{Ed} = W + (G + \psi_0 Q)\phi = 36 + (675 + 0.7 \times 542)/366 = 39 \text{ kN}$$

The lateral load applied to one edge of each floor is

$$h_{Ed} = 39/4 = 9.8 \text{ kN/m}$$

and for the 28-m span of the floor,

$$M_{Ed} = h_{Ed}L^2/8 = 9.8 \times 28^2/8 = 960 \text{ kN m}$$

Stresses and stiffness

Assuming a lever arm of about $0.8 \times 19 = 15.2$ m, the area of reinforcement needed near each edge of each floor slab, Fig. 5.21, is tiny:

$$A_s = (960/15.2)/0.435 = 145 \text{ mm}^2$$

The shear force applied to each shear wall is

$$9.8 \times 28/2 = 137 \text{ kN per storey}$$

The reinforced concrete wall HJ in Fig. 5.21 is a cantilever 36 m high. For storeys 4 m high, the horizontal load in its plane is

$$137/4 = 34.3 \text{ kN/m}$$

The bending moment at its base is

$$M_{Ed} = 34.3 \times 36^2/2 = 22\,230 \text{ kN m}$$

Elastic analysis for a beam 19 m deep and 200 mm wide gives the maximum bending stress as less than 2 N/mm^2. The in-plane deflection at the top of the wall, including shear deformation, and with $n = 2n_0 = 13.6$ (as for the columns), is less than 6 mm. This is an additional sidesway of $6/36\,000 = 1/6000$, which is less than 10% of the frame imperfection of $1/366$. The resulting increase in action effects is less than 10%, so the use of first-order global analysis is confirmed.

5.10   Example (continued): nominally-pinned joint to external column

The design vertical shear for this joint is 246 kN, from Equation 5.16. From standard details for partial-depth end-plate joints [51, 52], the initial design of the joint is as shown in Fig. 5.22, with six M20 8:8 bolts. The 220 mm × 150 mm end plate is of S275 steel and only 8 mm thick, so that its plastic deformation can provide the necessary end rotation for the beam while transmitting very little bending moment to the column.

Preliminary calculation showed that a four-bolt joint is just adequate for the vertical shear. However, it may be necessary to resist a tensile force of about 75 kN, depending on how the robustness of the structure is to be assured. Also, the greater depth of a six-bolt joint provides better torsional restraint to the beam during erection.

Detailed calculations to EN 1993-1-1 and EN 1993-1-8 are not given, as this is not a composite joint; but the results may be of interest. The calculated resistances are as follows:

- shear of six M20 bolts, $V_{Rd} = 565$ kN;
- bearing of bolts on end plate, $\Sigma F_{b,Rd} = 471$ kN;
- shear resistance of 6-mm fillet welds to beam web, $V_{w,Rd} = 377$ kN;
- shear resistance of 220-mm depth of beam web, $V_{Rd} = 767$ kN;
- block tearing failure of end plate, both sides on surfaces ABC (Fig. 5.22), $V_{Rd} = 412$ kN.

Thus, the resistance of the joint to vertical shear, neglecting bending moment and axial tension, is 377 kN, which provides sufficient margin for these other effects.

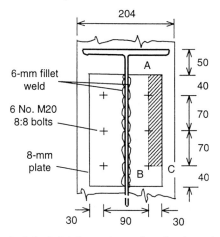

Figure 5.22   End-plate joint to major axis of external column

Appendix A
Partial-interaction theory

A.1 Theory for simply-supported beam

This subject is introduced in Section 2.6, which gives the assumptions and notation used in the theory that follows. On first reading, it may be found helpful to rewrite the algebraic work in a form applicable to a beam with the very simple cross-section shown in Fig. 2.2. This can be done by making these substitutions.

Replace A_c and A_a by bh, and d_c by h.
Replace I_c and I_a by $bh^3/12$.
Put $k_c = n = 1$, so that E'_c, E_c and E_s are replaced by E.

The beam to be analysed is shown in Fig. 2.15, and Fig. A.1 shows in elevation a short element of the beam, of length dx, distant x from the mid-span cross-section. For clarity, the two components are shown separated, and displacements are much exaggerated. The slip is s at cross-section x, and increases over the length of the element to $s + (ds/dx)\,dx$, which is written as s^+. This notation is used in Fig. A.1 for increments in the other variables, M_c, M_a, F, V_c and V_a, which are respectively the bending moments, axial force and vertical shears acting on the two components of the beam, the subscripts c and a indicating concrete and steel. It follows from longitudinal equilibrium that the forces F in steel and concrete are equal. The interface vertical force r per unit length is unknown, so it cannot be assumed that V_c equals V_a.

If the interface longitudinal shear is v_L per unit length, the force on each component is $v_L dx$. It must be in the direction shown, to be consistent with the sign of the slip, s. The load–slip relationship is

$$pv_L = ks \tag{A.1}$$

since the load per connector is pv_L.

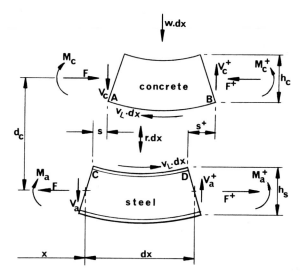

Figure A.1   Elevation of element of composite beam

We first obtain equations deduced from equilibrium, elasticity and compatibility, then eliminate M, F, V and v_L from them to obtain a differential equation relating s to x, and finally solve this equation and insert the boundary conditions. These are as follows.

(1)   Zero slip at mid-span, from symmetry, so

$$s = 0 \quad \text{when} \quad x = 0 \tag{A.2}$$

(2)   At the supports, M and F are zero, so the difference between the longitudinal strains at the interface is the differential strain, ε_c, and therefore

$$\frac{ds}{dx} = -\varepsilon_c \quad \text{when} \quad x = \pm \frac{L}{2} \tag{A.3}$$

Equilibrium
Resolve longitudinally for one component:

$$\frac{dF}{dx} = -v_L \tag{A.4}$$

Take moments:

$$\frac{dM_c}{dx} + V_c = \frac{1}{2}v_L h_c \quad \frac{dM_a}{dx} + V_a = \frac{1}{2}v_L h_s \tag{A.5}$$

The vertical shear at section x is wx, so

$$V_c + V_a = wx \tag{A.6}$$

Now $\frac{1}{2}(h_c + h_s) = d_c$, so from Equations A.5 and A.6,

$$\frac{dM_c}{dx} + \frac{dM_a}{dx} + wx = v_L d_c \tag{A.7}$$

Elasticity

In beams with adequate shear connection, the effects of uplift are negligible in the elastic range. If there is no gap between the two components, they must have the same curvature, ϕ, and simple beam theory gives the moment–curvature relations. Using Equation 2.19 for E'_c, then

$$\phi = \frac{M_a}{E_s I_a} = \frac{nM_c}{k_c E_a I_c} \tag{A.8}$$

The longitudinal strains in concrete along AB (Fig. A.1) and in steel along CD are:

$$\varepsilon_{AB} = \frac{1}{2}h_c\phi - \frac{nF}{k_c E_a A_c} - \varepsilon_c \tag{A.9}$$

$$\varepsilon_{CD} = -\frac{1}{2}h_s\phi + \frac{F}{E_a A_a} \tag{A.10}$$

where ε_c is the free shrinkage strain of the concrete, taken as positive.

Compatibility

The difference between ε_{AB} and ε_{CD} is the slip strain, so from Equations A.9 and A.10, and putting $\frac{1}{2}(h_c + h_s) = d_c$,

$$\frac{ds}{dx} = \phi d_c - \frac{F}{E_a}\left(\frac{n}{k_c A_c} + \frac{1}{A_a}\right) - \varepsilon_c \tag{A.11}$$

It is now possible to derive the differential equation for s. Eliminating M_c and M_a from Equations A.7 and A.8,

$$E_a\left(\frac{k_c I_c}{n} + I_a\right)\frac{d\phi}{dx} + wx = v_L d_c \tag{A.12}$$

From Equations A.1 and 2.22,

$$\frac{d\phi}{dx} = \frac{kd_c s/p - wx}{E_a I_0} \tag{A.13}$$

Differentiating Equation A.11 and eliminating ϕ from Equation A.13, F from Equation A.4, and v_L from Equation A.1:

$$\frac{d^2 s}{dx^2} = \frac{kd_c^2 s/p - wd_c x}{E_a I_0} + \frac{ks}{E_a A_0 p} = \frac{ks}{pE_a I_0}\left(d_c^2 + \frac{I_0}{A_0}\right) - \frac{wd_c x}{E_a I_0}$$

Introducing A' from Equation 2.21, α^2 from Equation 2.23 and β from Equation 2.24 gives Result 2.25, which is in a standard form:

$$\frac{d^2 s}{dx^2} - \alpha^2 s = -\alpha^2 \beta wx \tag{2.25}$$

Solving for s,

$$s = K_1 \sinh \alpha x + K_2 \cosh \alpha x + \beta wx \tag{A.14}$$

The boundary conditions, Equations A.2 and A.3, give

$$K_2 = 0 \quad \varepsilon_c = -K_1 \alpha \cosh(\alpha L/2) - \beta w$$

and substitution in Equation A.14 gives s in terms of x:

$$s = \beta wx - \left(\frac{\beta w + \varepsilon_c}{\alpha}\right) \text{sech}\left(\frac{\alpha L}{2}\right) \sinh \alpha x \tag{2.27}$$

Other results can now be found as required. For example, the slip strain at mid-span is

$$\left(\frac{ds}{dx}\right)_{x=0} = \beta w - (\beta w + \varepsilon_c)\,\text{sech}(\alpha L/2) \tag{A.15}$$

and the slip at $x = L/2$ due to ε_c alone (i.e. with $w = 0$), is

$$(s)_{x=L/2} = -\frac{\varepsilon_c}{\alpha} \tanh\left(\frac{\alpha L}{2}\right) \tag{A.16}$$

A.2   Example: partial interaction

These calculations are introduced in Section 2.7. They relate to a beam shown in section in Fig. 2.16, which carries a distributed load w per unit length over a simply-supported span L. The materials are assumed to be concrete with a characteristic cube strength of 30 N/mm^2 and mild steel, with a characteristic yield strength of 250 N/mm^2. Creep is neglected ($k_c = 1$) and we assume $n = 10$, so for the concrete $E_c = E'_c = 20$ kN/mm^2, from Equation 2.19.

The dimensions of the beam (Fig. 2.16) are so chosen that the transformed cross-section is square: $L = 10$ m, $b = 0.6$ m, $h_c = h_s = 0.3$ m. The steel member is thus a rectangle of breadth 0.06 m, so that $A_a = 0.018$ m^2, $I_a = 1.35 \times 10^{-4}$ m^4.

The design of' such a beam on an ultimate-strength basis is likely to lead to a working or 'service' load of about 35 kN/m. If stud connectors 19 mm in diameter and 100 mm long are used in a single row, an appropriate spacing would be 0.18 m. Push-out tests give the ultimate shear strength of such a connector as about 100 kN, and the slip at half this load is usually between 0.2 and 0.4 mm. Connectors are found to be stiffer in beams than in push-out tests, so a connector modulus $k = 150$ kN/mm will be assumed here, corresponding to a slip of 0.33 mm at a load of 50 kN per connector.

The distribution of slip along the beam and the stresses and curvature at mid-span are now found by partial-interaction theory, using the results obtained in Section A.1, and also by full-interaction theory. The results are discussed in Section 2.7.

First α and β are calculated. From Equation 2.22 with $I_c = nI_a$ (from the shape of the transformed section) and $k_c = 1$, $I_0 = 2.7 \times 10^{-4}$ m^4.

From Equation 2.20 with $A_c = nA_a$ and $k_c = 1$, $A_0 = 0.009$ m^2.
From Equation 2.21, $1/A' = 0.3^2 + (2.7 \times 10^{-4})/0.009 = 0.12$ m^2.
From Equation 2.23, with $k = 150$ kN/mm and $p = 0.18$ m,

$$\alpha^2 = \frac{150 \times 0.12}{0.18 \times 200 \times 0.27} = 1.85 \text{ m}^{-2}$$

whence $\alpha = 1.36$ m^{-1}. Now $L = 10$ m, so $\alpha L/2 = 6.8$ and sech$(\alpha L/2) = 0.002\,23$. From Equation 2.24,

$$\beta = \frac{0.18 \times 0.3}{0.12 \times 150 \times 1000} = 3.0 \times 10^{-6} \text{ m/kN}$$

We assumed $w = 35$ kN/m, so $\beta w = 1.05 \times 10^{-4}$ and $\beta w/\alpha = 0.772 \times 10^{-4}$ m. An expression for the slip in terms of x is now given by Equation 2.27 with $\varepsilon_c = 0$:

$$10^4 s = 1.05x - 0.0017 \sinh(1.36x) \tag{2.28}$$

This gives the maximum slip (when $x = \pm 5$ m) as ± 0.45 mm.

This may be compared with the maximum slip if there were no shear connection, which is given by Equation 2.6 as

$$\frac{wL^3}{4Ebh^2} = \frac{35 \times 10^3}{4 \times 20 \times 0.6 \times 0.3^2 \times 1000} = 8.1 \text{ mm}$$

The stresses at mid-span can be deduced from the slip strain and the curvature. Differentiating Equation 2.28 and putting $x = 0$,

$$10^4 \left(\frac{ds}{dx} \right)_{x=0} = 1.05 - 0.0017 \times 1.36 = 1.05$$

so the slip strain at mid-span is 105×10^{-6}. From Equation A.13,

$$\frac{d\phi}{dx} = 4.64s - 6.5 \times 10^{-4}x$$

Using Equation 2.28 for s and integrating,

$$10^6 \phi = -81.5x^2 - 0.585 \cosh(1.36x) + K$$

The constant K is found by putting $\phi = 0$ when $x = L/2$, whence at $x = 0$,

$$\phi = 0.0023 \text{ m}^{-1}$$

The corresponding change of strain between the top and bottom faces of a member 0.3 m deep is 0.3×0.0023, or 690×10^{-6}. The transformed cross-section is symmetrical about the interface, so the strain in each material at this level is half the slip strain, say 52×10^{-6}, and the strain distribution is as shown in Fig. 2.17. The stresses in the concrete, found by multiplying the strains by E_c (20 kN/mm^2), are 1.04 N/mm^2 tension and 12.8 N/mm^2 compression. The tensile stress is below the cracking stress, as assumed in the analysis.

The maximum compressive stress in the concrete is given by full-interaction theory (Equation 2.7) as

$$\sigma_{cf} = \frac{3wL^2}{16bh^2} = \frac{3 \times 35 \times 100}{16 \times 0.6 \times 0.09 \times 10^3} = 12.2 \text{ N/mm}^2$$

References

1. Fisher, J.W. (1970) Design of composite beams with formed metal deck. *Eng. J. Amer. Inst. Steel Constr.*, **7**, July, 88–96.
2. Wang, Y.C. (2002) *Steel and Composite Structures – Analysis and Design for Fire Safety.* Spon, London.
3. British Standards Institution BS EN 1994. *Design of composite steel and concrete structures. Part 1-1, General rules and rules for buildings.* To be published, British Standards Institution, London.
4. Kerensky, O.A. & Dallard, N.J. (1968) The four-level interchange between M4 and M5 motorways at Almondsbury. *Proc. Instn Civ. Engrs*, **40**, July, 295–322.
5. Johnson, R.P., Finlinson, J.C.H. & Heyman, J. (1965) A plastic composite design. *Proc. Instn Civ. Engrs*, **32**, Oct, 198–209.
6. Cassell, A.C., Chapman, J.C. & Sparkes, S.R. (1966) Observed behaviour of a building of composite steel and concrete construction. *Proc. Instn Civ. Engrs*, **33**, April, 637–658.
7. Couchman, G. & Way, A. (1998) *Joints in steel construction – Composite connections, Publication 213*, Steel Construction Institute, Ascot.
8. Anderson, D., Aribert, J-M., Bode, H. & Kronenburger, H.J. (2000) Design rotation capacity of composite joints. *The Structural Engineer*, **78**, 6, 25–29, London.
9. Brown, N.D. & Anderson, D. (2001) Structural properties of composite major-axis end plate connections. *J. Constr. Steel Research*, **57**, 327–349.
10. British Standards Institution BS EN 1993. *Design of steel structures. Part 1-8: Design of joints.* To be published, British Standards Institution, London.
11. European Convention for Constructional Steelwork (1981) *Composite Structures.* The Construction Press, London.
12. British Standards Institution (2002) BS EN 1990. *Eurocode: Basis of structural design.* British Standards Institution, London.
13. British Standards Institution (2002) BS EN 1991. *Actions on structures. Part 1-1: General actions – Densities, self weight, imposed*

loads for buildings. British Standards Institution, London. (Parts 1-2, 1-3, etc. specify other types of action.)

14. British Standards Institution BS EN 1992. *Design of concrete structures. Part 1-1, General rules and rules for buildings.* To be published, British Standards Institution, London.

15. British Standards Institution BS EN 1993. *Design of steel structures. Part 1-1, General rules and rules for buildings.* To be published, British Standards Institution, London, with 18 other Parts.

16. British Standards Institution BS EN 1994. *Design of composite steel and concrete structures. Part 1-2, Structural fire design.* To be published, British Standards Institution, London.

17. Johnson, R.P. & Anderson, D. (2004) *Designers' Guide to EN 1994-1-1: Eurocode 4, Design of Composite Steel and Concrete Structures. Part 1-1, General rules and rules for buildings.* Thomas Telford, London.

18. Beeby, A.W. & Narayanan, R.S. *Designers' Guide to EN 1992-1-1: Eurocode 2, Design of concrete structures. Part 1-1, General rules and rules for buildings.* To be published, Thomas Telford, London.

19. British Standards Institution (1990) BS 5950. Part 3, Section 3.1: *Code of practice for design of simple and continuous composite beams* and (1994) Part 4: *Code of practice for design of floors with profiled steel sheeting.* British Standards Institution, London.

20. British Standards Institution BS EN 10080. *Steel for the reinforcement of concrete.* To be published, British Standards Institution, London.

21. British Standards Institution (1990) BS EN 10025. *Hot rolled products of non-alloy structural steels and their technical delivery conditions.* British Standards Institution, London.

22. British Standards Institution (1997) BS 8110. *Structural use of concrete. Part 1: Code of practice for design and construction.* British Standards Institution, London.

23. Goble, G.G. (1968) Shear strength of thin-flange composite specimens, *Eng. J. Amer. Inst. Steel Constr.*, **5**, April, 62–65.

24. Mottram, J.T. & Johnson, R.P. (1990) Push tests on studs welded through profiled steel sheeting. *The Structural Engineer*, **68**, 15 May, 187–193.

25. Johnson, R.P., Greenwood, R.D. & van Dalen, K. (1969) Stud shear-connectors in hogging moment regions of composite beams. *The Structural Engineer*, **47**, Sept, 345–350.

26. British Standards Institution (1967) CP 117, *Composite construction in structural steel and concrete.* Part 2, *Beams for bridges.* British Standards Institution, London.

27. Patrick, M. & Bridge, R.Q. (1993) Design of composite slabs for vertical shear. In *Composite construction in steel and concrete II* (Easterling, W.S. & Roddis, W.M.K., eds), Proc. Engineering Foundation Conference, Amer. Soc. Civ. Engrs, New York.

28. Johnson, R.P. The *m–k* and partial-interaction models for shear resistance of composite slabs, and the use of non-standard test data. To be published in: *Composite construction in steel and concrete V*, Proceedings of a Conference, Kruger National Park, 2004. Amer. Soc. Civ. Engrs, New York.

29. Bode, H. & Sauerborn, I. (1993) Modern design concept for composite slabs with ductile behaviour. In *Composite construction in steel and concrete II* (Easterling, W.S. & Roddis, W.M.K., eds), Proc. Engineering Foundation Conference, Amer. Soc. Civ. Engrs, New York.

30. Patrick, M. & Bridge, R.Q. (1990) Parameters affecting the design and behaviour of composite slabs. Symposium, Mixed structures, including new materials, Brussels. *Reports, Int. Assoc. for Bridge and Struct. Engrg*, **60**, 221–225.

31. Slutter, G.C. & Driscoll, G.C. (1965) Flexural strength of steel–concrete composite beams. *Proc. Amer. Soc. Civ. Engrs*, **91**, ST2, April, 71–99.

32. Chung, K.F. & Lawson, R.M. (2001) Simplified design of composite beams with large web openings to Eurocode 4. *J. Constr. Steel Research*, **57**, Feb., 135–164.

33. Johnson, R.P. & Molenstra, N. (1991) Partial shear connection in composite beams for buildings. *Proc. Instn Civ. Engrs, Part 2*, **91**, Dec., 679–704.

34. Beeby, A.W. (1971) The prediction and control of flexural cracking in reinforced concrete members. *Publication SP 30*, 55–75. American Concrete Institute, Detroit.

35. British Standards Institution (1992) BS 6472. *Guide to evaluation of human exposure to vibration in buildings*. British Standards Institution, London.

36. Wyatt, T.A. (1989) *Design guide on the vibration of floors. Publication 076*, Steel Construction Institute, Ascot.

37. Lawson, R.M., Bode, H., Brekelmans, J.W.P.M., Wright, P.J. & Mullett, D.L. (1999) 'Slimflor' and 'Slimdek' construction: European developments. *The Structural Engineer*, **77**, 20 April, 22–30.

38. Hicks, S. & Lawson, R.M. (2003) *Design of composite beams using pre-cast concrete slabs. Publication 287*, Steel Construction Institute, Ascot.

39. Allison, R.W., Johnson, R.P. & May, I.M. (1982) Tension field action in composite plate girders. *Proc. Instn Civ. Engrs, Part 2*, **73**, June, 255–276.

40. Johnson, R.P. & Chen, S. (1991) Local buckling and moment redistribution in Class 2 composite beams. *Struct. Engrg International*, **1**, Nov., 27–34.

41. Lawson, R.M. & Rackham, J.W. (1989) *Design of haunched composite beams in buildings. Publication 060*, Steel Construction Institute, Ascot.

42.   Johnson, R.P. & Allison, R.W. (1983) Cracking in concrete tension flanges of composite T-beams. *The Structural Engineer*, **61B**, March, 9–16.

43.   Hensman, J.S. & Nethercot, D.A. (2001) Design of unbraced composite frames using the wind-moment method. *The Structural Engineer*, **79**, 5 June, 28–32.

44.   Faber, O. (1956) More rational design of cased stanchions. *The Structural Engineer*, **34**, March, 88–109.

45.   Jones, R. & Rizk, A.A. An investigation on the behaviour of encased steel columns under load. *The Structural Engineer*, **41**, Jan., 21–33.

46.   Basu, A.K. & Sommerville, W. (1969) Derivation of formulae for the design of rectangular composite columns. *Proc. Instn Civ. Engrs, Supp. vol.*, 233–280.

47.   Virdi, K.S. & Dowling, P.J. (1973) The ultimate strength of composite columns in biaxial bending. *Proc. Instn Civ. Engrs, Part 2*, **55**, March, 251–272.

48.   Johnson, R.P. & May, I.M. (1978) Tests on restrained composite columns. *The Structural Engineer*, **56B**, June, 21–28.

49.   Neogi, P.K., Sen, H.K. & Chapman, J.C. (1969) Concrete-filled tubular steel columns under eccentric loading. *The Structural Engineer*, **47**, May, 187–195.

50.   Coates, R.C., Coutie, M.G. & Kong, F.K. (1988) *Structural Analysis*, 3rd edn. Van Nostrand Reinhold (UK), Wokingham.

51.   Steel Construction Institute (1993) *Joints in simple construction. Vol. 1: Design methods*, 2nd edn. *Publication 205*, Steel Construction Institute, Ascot.

52.   Steel Construction Institute (1992) *Joints in simple construction. Vol. 2: Practical applications. Publication 206*, Steel Construction Institute, Ascot.

Index